Secure Communications

WILEY SERIES IN COMMUNICATIONS NETWORKING & DISTRIBUTED SYSTEMS.

Series Editor: David Hutchison, Lancaster University
Series Advisers: Harmen van As, TU Vienna
                   Serge Fdida, University of Paris
                   Joe Sventek, Agilent Laboratories, Edinburgh

The 'Wiley Series in Communications Networking & Distributed Systems' is a series of expert-level, technically detailed books covering cutting-edge research and brand new developments in networking, middleware and software technologies for communications and distributed systems. The books will provide timely, accurate and reliable information about the state-of-the-art to researchers and development engineers in the Telecommunications and Computing sectors.

Other titles in the series:

Wright: Voice over Packet Networks
Jepsen: Java in Telecommunications

Secure Communications
Applications and Management

Roger J. Sutton
Crypto AG, Switzerland

JOHN WILEY & SONS, LTD

Other Wiley Editorial Offices

John Wiley & Sons, Inc., 605 Third Avenue,
New York, NY 10158-0012, USA

WILEY-VCH Verlag GmbH
Pappelallee 3, D-69469 Weinheim, Germany

John Wiley & Sons Australia Ltd, 33 Park Road, Milton,
Queensland 4064, Australia

John Wiley & Sons (Canada) Ltd, 22 Worcester Road
Rexdale, Ontario, M9W 1L1, Canada

John Wiley & Sons (Asia) Pte Ltd, 2 Clementi Loop #02-01,
Jin Xing Distripark, Singapore 129809

British Library Cataloguing in Publication Data

A catalogue record for this book is available from the British Library

ISBN 0471 49904 8

Typeset in Times by Deerpark Publishing Services Ltd, Shannon, Ireland.

This book is printed on acid-free paper responsibly manufactured from sustainable forestry, in which at least two trees are planted for each one used for paper production.

Dedication

To my family without whose support, advice and patience, I would not have had the stamina and discipline to see this production through. To my mother, Margaret Jean Sutton who as a member of the ATS during the Second World War, played a small role in the securing of Britain's communications.

To those whom I have loved and those whom, for some obscure reason and known only to themselves, have seen fit to hold me in their affections.

To those people of many nations, language, race or creed with whom I have shared friendships that have broken through all manmade barriers and who have made my journeys so delightfully rewarding.

Contents

Preface   xiii

Acknowledgement   xv

Glossary   xix

Acronyms and Abbreviations   xxiii

1   Threats and Solutions   **1**
1.1   The Technical Threats to Communications Security   4
1.2   Authentication   4
    1.2.1   Text/Data Message Authentication   6
1.3   Confidentiality   7
1.4   Integrity   8
    1.4.1   Digital Signatures   8
1.5   Availability   10
    1.5.1   PINs and Passwords   10
    1.5.2   Biometric Access Tools   13
    1.5.3   Challenge/Response Control   14
    1.5.4   Tamperproof Modules   15
1.6   Compromising Emanation/Tempest Threats   16
    1.6.1   Compromising Emanation Definitions   16
    1.6.2   Compromising Emanation   16
    1.6.3   Modulated Harmonics   16
    1.6.4   Electronic Coupling   19
    1.6.5   Preventative Measures in Electronic Equipment Construction   21

2   An Introduction to Encryption and Security Management   **25**
2.1   Analogue Scrambling   25
    2.1.1   Phonemes and the Structure of Voice Signals   26
    2.1.2   Frequency Scrambling   28
    2.1.3   Time Element Scrambling   29
    2.1.4   Digital Ciphering   30
    2.1.5   Digital Stream Ciphering   31
    2.1.6   Block Ciphering   33
    2.1.7   Summary   37
2.2   Algorithms   38
    2.2.1   Symmetrical Cryptography   38
    2.2.2   Asymmetrical Cryptography   39
    2.2.3   Hash Algorithms   41
    2.2.4   MACs (Message Authentication Codes)   41

	2.2.5	Digital Signature Algorithms	41
	2.2.6	Key Agreement/Exchange Algorithms	42
	2.2.7	Summary of Comparisons Between Asymmetric and Symmetric Algorithms	42
2.3	Goodbye DES, Hello AES		43
2.4	Fundamentals in Key Management		44
	2.4.1	Key Generation	45
	2.4.2	Key Storage	47
	2.4.3	Key Distribution	48
	2.4.4	Key Changes	51
	2.4.5	Key Destruction	54
	2.4.6	Separation	55
2.5	Evaluating Encryption Equipment		57
	2.5.1	The Main Points of Evaluation	58

3	**Voice Security in Military Applications**		**61**
3.1	Analogue Encryption of Naval Long Range, HF Radio Communications		62
	3.1.1	Ship Communications Operation	63
	3.1.2	The Cipher/Scrambler Features	65
	3.1.3	Synchronisation	68
	3.1.4	Security Parameters	69
	3.1.5	Key Distribution and Management	70
3.2	Stand-alone Digital Cipher Units in Land-based Operations		70
	3.2.1	The Ground Force Scenario	71
	3.2.2	The Cipher Unit Features	71
	3.2.3	Synchronisation	74
	3.2.4	Security Parameters	76
	3.2.5	Key Management	77
3.3	Radio Integrated Cipher Module		82
	3.3.1	Typical Features	83
	3.3.2	Cryptographic Parameters	83
	3.3.3	Other Security Parameters and Features	83

4	**Telephone Security**		**87**
4.1	Specific Threats to Telephone Operations		88
	4.1.1	Telephone Security Requirements and Features	89
4.2	Network Technologies		90
	4.2.1	Secure Telephone Communication	90
	4.2.2	INMARSAT Communications	91
4.3	Telephone Security Solutions		95
	4.3.1	STU III/IIB	96
	4.3.2	The Alternative Telephone Security	98
	4.3.3	Hardware Security Features	101
	4.3.4	Telephone Security Architecture and Functions	102
4.4	Key and Access Management		103
	4.4.1	The Complete Key System	104
	4.4.2	The 'ZUPPA' Network	107
4.5	Network Implementation		108
4.6	Key Distribution		111
4.7	Summary		112

5	**Secure GSM Systems**		**113**
5.1	The Basic GSM Architecture		113
	5.1.1	System Components	114
	5.1.2	The GSM Subsystems	116

	5.1.3	The GSM Radio Um Interface	117
5.2	Standard GSM Security Features		118
	5.2.1	The AuC	119
	5.2.2	The HLR	119
	5.2.3	The VLR	120
	5.2.4	SIM Card	120
	5.2.5	The IMSI & TMSI	121
	5.2.6	Standard GSM Encryption	121
	5.2.7	Cryptographic Attacks on the GSM Algorithms	126
	5.2.8	TDMA Time Division Multiple Access	127
	5.2.9	Frequency Hopping	127
5.3	Custom Security for GSM Users		128
	5.3.1	The Custom Encryption Process	130
	5.3.2	Key Systems	133
	5.3.3	Cryptographic Parameters and Algorithms	135
	5.3.4	Security Architecture	135
	5.3.5	Cipher Unit Hardware Elements	136
	5.3.6	System Overview with Secure GSM and Fixed Subscriber Equipment	137
5.4	Key Management and Tools		138
	5.4.1	Key Distribution and Loading	138
	5.4.2	Chip Cards and Readers	138
	5.4.3	Key Signatures	138
5.5	GPRS General Packet Radio Systems		139
	5.5.1	Basic GPRS Operation and Security	139

6	**Security in Private VHF/UHF Radio Networks**		**143**
6.1	Applications and Features		143
	6.1.1	The Ship Group	143
	6.1.2	The Escort Group	145
	6.1.3	The Close Support Group	145
	6.1.4	The Telephone Groups	146
6.2	Threats		146
	6.2.1	Confidentiality	146
	6.2.2	Integrity	146
	6.2.3	Authenticity	147
	6.2.4	Access	147
6.3	Countermeasures		147
	6.3.1	Protection of Confidentiality	147
	6.3.2	Authentication	147
	6.3.3	Access Control	149
6.4	Communications Network Design and Architecture		150
	6.4.1	The Close Support Group	151
	6.4.2	The Escort Group	152
	6.4.3	The Ship Group	152
6.5	Hardware Components and Functions		153
	6.5.1	Hand-held UHF Radios	153
	6.5.2	Base Stations/Repeaters	158
	6.5.3	Telephone Patch	160
	6.5.4	Security Management Tools	162
6.6	Security and Key Management		162
	6.6.1	Functions of the Management Centre	162
	6.6.2	Frequency Management	163
	6.6.3	Key Management	165
6.7	Other Security Features		168
	6.7.1	Remote Key Cancelling	168

| | 6.7.2 | Remote Blocking | 168 |
| | 6.7.3 | Silent Mode Tracking | 168 |

7   Electronic Protection Measures – Frequency Hopping   171

7.1	ESM		171
7.2	EA		172
7.3	EPM		172
	7.3.1	Methods of Attack	172
	7.3.2	Spread Spectrum Techniques	177
	7.3.3	COMSEC and TRANSEC	182
7.4	Military Applications		183
	7.4.1	Applications Requirements	183
	7.4.2	Operational Requirements	184
	7.4.3	Security Requirements	184
	7.4.4	Anti Jamming Requirements	184
	7.4.5	Co-location	185
	7.4.6	Air Defence Scenario	185
	7.4.7	Close Air Support Scenario	187
7.5	Network Architecture and Management		189
	7.5.1	Mission Procedures	190
7.6	Characteristics of Frequency Hopping Networks		191
	7.6.1	COMSEC	191
	7.6.2	TRANSEC	192
7.7	Key/Data Management and Tools		201
	7.7.1	Algorithm Data	202
	7.7.2	Frequency Data	203
	7.7.3	Pre-set Data	203
	7.7.4	Configuration Parameters	203
	7.7.5	Key Distribution	203
	7.7.6	The Time Problem	206
7.8	Hardware Components		207
	7.8.1	Airborne Transceiver	207

8   Link and Bulk Encryption   211

8.1	Basic Technology of Link Encryption		211
	8.1.1	Frame Modes	211
8.2	The Ciphering Process		212
8.3	Cryptographic Parameters		214
	8.3.1	Key Agreement	216
8.4	Key and Network Management		216
	8.4.1	Civilian Application	216
8.5	Military Link Security		222
	8.5.1	Military Topology and Features	223

9   Secure Fax Networks   227

9.1	Basic Facsimile Technology		228
9.2	The Basic Operation of an Encrypted Fax Machine		229
	9.2.1	Fax by Telephone Line	229
	9.2.2	Fax by Radio	230
	9.2.3	The GB Fax Protocol	230
9.3	Manual/Automatic Key Selection		232
	9.3.1	Multi-key Fax Networks	233
	9.3.2	Single Key Fax Networks	234
	9.3.3	Facsimile Transmission over Radio	235

9.4 Network Architecture 235
    9.4.1 Interesting Features of DEFNET 236
    9.4.2 Ministry of Defence Subnet 237
9.5 Key Management and Tools 237
    9.5.1 Key Management of DEFNET 237
    9.5.2 Key Generation 237
    9.5.3 Operating Parameters 239
    9.5.4 Key and Parameter Distribution 239
9.6 Fax Over Satellite Links 239

10 PC Security **241**
10.1 Security Threats and Risks 244
10.2 Implementation of Solutions 244
    10.2.1 Unauthorised Read Out of Data Stored on Local Storage Media 244
    10.2.2 Unauthorised Read Out of 'Deleted' Data 245
    10.2.3 Unauthorised Read Out of Data Stored on a Remote LAN 246
    10.2.4 Unauthorised Manipulation of Data Stored on a LAN 247
    10.2.5 Eavesdropping on an Untrusted LAN or Public Network 249
    10.2.6 Spoofing or Masquerading 249
    10.2.7 Unauthorised Manipulation of Data During Transmission over a Public Network 249
    10.2.8 Unauthorised Access to/Read Out of/Analysis of/Manipulation of/the Security System 249
    10.2.9 'Brute-force' Attack 250
    10.2.10 Inefficient Security and Key Management 251
    10.2.11 Analysis of Residual Plain Information 251
    10.2.12 The Compromise of Information, Due to Loss or Theft of Equipment or the Transfer of Security Personnel 251
    10.2.13 The Storage or Transmission of Data in Plain, Due to Loss of Keys or Key Incompatibility 252
    10.2.14 Illegal Access to Equipment Under Maintenance or Repair 252
    10.2.15 Unauthorised Intrusion into the PC Environment Whilst Connected to a Public or Untrusted Network 252
10.3 Access Protection 253
    10.3.1 Access Control Systems 253
    10.3.2 Access by Chip Card 254
    10.3.3 Access by PC Cards 254
    10.3.4 Access by PCMCIA Module 256
10.4 Boot-up Protection by On-board Hardware with Smart Card 256
10.5 LAN Security 256
    10.5.1 LAN Workstation Scenario 257
    10.5.2 Business Trip Notebook Scenario 258
10.6 Model Application of PC Security 259
10.7 System Administration 264

11 Secure E-mail **265**
11.1 The E-mail Scenario 265
11.2 Threats 267
    11.2.1 Information Disclosure 267
    11.2.2 Modification of Messages 269
    11.2.3 Replay Attack 269
    11.2.4 Masquerading 269
    11.2.5 Spoofing 269
    11.2.6 Denial of Service Attack 269
11.3 Type and Motivation of Attackers 270
11.4 Methods of Attack 270

11.5   Countermeasures   271
11.6   Guidelines for E-mail Security   274

12   Secure Virtual Private Networks   **275**
12.1   Scenario   275
12.2   Definition of VPN   275
12.3   Protocols   277
12.4   Packet Header Formats   278
12.5   Security Association List   281
12.6   Tunnel Table   282
12.7   Routing Tables   282
12.8   Packet Filtering   283
12.9   Threats and Countermeasures   284
         12.9.1   *Attacks Within the Public Network*   285
         12.9.2   *Attacks Within Nodes of the Trusted Network*   285
         12.9.3   *Attacks Aimed at Gaining Access to the Private Network*   285
12.10 Example Application – 'Diplomatic Network'   285

13   Military Data Communications   **289**
13.1   Applications   290
         13.1.1   *Data Over Radio Links*   290
         13.1.2   *Modes of Radio Operation, Automatic Repeat Request and Forward Error*
                      Correction   290
         13.1.3   *Use of GAN Terminals in Battlefield Applications*   291
13.2   Data Terminals and Their Operating Features   292
13.3   Technical Parameters   293
13.4   Security Management   294
         13.4.1   *Access Control*   294
         13.4.2   *Data Encryption*   295
         13.4.3   *Loss of Data*   295
         13.4.4   *TEMPEST*   295
13.5   Key Management   295
13.6   Combat Packet Data Networks   296
         13.6.1   *Packet Radios*   296
         13.6.2   *Packet Data Networks*   297

14   Management, Support and Training   **301**
14.1   Environments of Security Management   303
         14.1.1   *The Global Environment*   303
         14.1.2   *The Local/Task Environment*   306
14.2   Infrastructure and Planning   307
         14.2.1   *Strategic Goals*   308
         14.2.2   *Tactical Goals*   308
         14.2.3   *Operational Goals*   308
14.3   Operational Hierarchies   308
14.4   Training   310
14.5   Customer Support   312
14.6   Troubleshooting   312
         14.6.1   *The Scanning Stage*   312
         14.6.2   *The Categorisation Stage*   313
         14.6.3   *The Diagnostics Stage*   313
         14.6.4   *Generating Solutions*   313

References   **315**

Index   **317**

Preface

This is not a book about cryptography, it is about how to apply cryptography to secure telecommunications. There are many fine manuscripts written on the subjects of cryptography and of telecommunications, but few that address the practical links between the two. In the eyes of the Cryptographer and in the ideals of those who employ him, security is often a matter of algorithms and mathematical statistics. Yet this is just the tip of the iceberg and what lies beneath this 'pristine peak' is a domain that is full of circumstance and danger. It is this grey area that I have addressed in this book and is written on the back of fifteen years experience in applying cryptography, technological know-how and psychological persuasion to the securing of my client's communications. In my experience the weak links in security have not necessarily been the strength of algorithms and hardware but rather in the way and diligence, or lack of it, that these have been implemented. As a result, what is the actual state of security at the communications level of an organisation is often very far removed from the grand ideals of the strategic decision makers. This book is aimed at providing a warning to those with their heads in the clouds and providing guidance to those who are given the task of implementing their strategies.

Secure Communications is essentially written in two parts and although some distinction is drawn between the technologies of voice and data communications, the two components are really a) The technical and philosophical aspects of security, support of chapters one, two and fourteen and b) The application chapters. The supporting chapters are included to provide background preparation material to the less cryptographically experienced reader. The application chapters also provide a varying degree of technical support specific to the medium in question, for without some knowledge of the medium technologies, it is impossible to assess the strengths and weaknesses of that technology, with any confidence.

Whilst there are many common factors between the applications, I have tried to view the problems of each platform from a different point of view. There are two reasons for my adopting this approach. The first is that it would be difficult to address each communications technology in any depth without having the security aspects readily at hand. This convenience is at the cost of some repetition that would be apparent to any cover-to-cover reader. The second reason being that each technology and each application can present many different approaches to securing them and to mass these into a single chapter would be too demanding and perhaps tedious for many a reader. Therefore, whilst I have striven to offer more complete platform packages in the application chapters, reading through the whole book should present a comprehensive package of alternative solutions. Essentially, I have sought to stimulate thought on the weaknesses of various communication technologies and present the strengths of a selection of solutions. The security manager reading this manuscript is expected to carry out something of a cut and paste exercise in applying the solutions suggested here, to secure his specific application.

One of the difficulties encountered in writing a book about security is the gaining of access to useful material and the freedom to publish it. Manufacturers and clients understandably strive to maintain their security and as a result, there have been times when I have experienced difficulty in acquiring material and permission to publish it. This is the nature of the industry and the reader, like the author, has to accept it. There are also occasions when the purists might argue that detail has been lost in some of the modelling that I have adopted. Bearing in mind the target audience of the book, these times were when I felt that the general concept was more important to portray rather than the delving into specific and complex issues.

Disclaimer

The author would like to point out that the opinions expressed in this text are solely those of his own and not of his employer, their agents or clients.

Acknowledgement

In writing this book, I am deeply indebted to my friend, Peter Mash for his writing of Chapter 12 and his many other contributions throughout. To Ralph Bühler, Dr. David Callaghan, Torgrim Jorgensen, Harry Kernohan, Dr. Richard Weber and others, for their technical criticisms and most valuable advice, I give my sincere thanks. Without their support, I could not have entertained writing this book. To Elisabeth for her advice and patience and meticulousness in proof reading the text, I offer my admiration and gratitude for tackling an arduous task.

Roger J. Sutton
106112.1717@compuserve.com

Figures 2.2, 3.6, 4.2, 5.10, 5.11, 5.18, 5.19, 6.5, 10.6 and 14.3 were reproduced with the kind permission of Crypto AG, Switzerland.

44 33 20 21 40 42 20 40 52 21 51 20 40 40 43 45 40 43

52 33 33 54 21 12 51 50 43 32 22 21 52 41 40 31 40 34

52 44 20 33 32 42 53 55 54 43 20 40 50 51 50 43 44 20

33 32 44 33 33 54 21 13 30 33 53 50 43 20 10 43 40 50

13

Glossary

Advanced encryption standard (AES): The replacement algorithm for DES, produced by Vincent Rijman and Joan Daemen.

Algorithm: A cryptographic procedure that defines how ciphering/deciphering is carried out.

Asymmetric algorithm: A cryptographic algorithm that uses different keys for encryption and decryption.

Authentication: The process of verifying that a particular name belongs to a particular entity.

Biometric access: The science of applying biological characteristics of a user as access tokens to a device or system, e.g. fingerprints.

Black designation: A designation given to cables, components, equipment and systems, which carry un-classified signals.

Block cipher: A cipher that encrypts data in blocks of a fixed size.

Brute force attack, exhaustive key search: The process of trying to recover a key or password by trying all the possibilities.

Certificate, public key: A specially formatted block of data that contains a public key and the owner's identification. The certificate carries the digital signature of a certifying authority to authenticate it.

Cipher: A procedure that transforms data from plaintext to ciphertext.

Cipher block chaining (CBC): A block mode cipher that combines the previous block of ciphertext with the current block of plaintext before encrypting it.

Cipher feedback (CFB): A block cipher mode that feeds previously encrypted ciphertext through the block cipher to generate the key that encrypts the next block of ciphertext.

Ciphertext: Data that have been encrypted by a cipher.

Compromising emanations: The radiation of electromagnet signals that can carry unintentionally, information about data within the system.

Confidentiality: The ability to ensure that information is not disclosed to persons who are not explicitly intended to read it.

Cryptanalysis: The process of trying to recover secret keys, or text from a ciphertext.

Cryptography: Mechanisms used to protect information by applying transformations to plaintext that are difficult to reverse without possessing knowledge of that mechanism.

Data encryption standard (DES): A block cipher that uses a key length of 56 bits, which is widely used in commercial systems.

Decipher; decrypt: Change from ciphertext into plaintext.

Diffie–Hellman (DH): A public key cipher algorithm that generates a shared secret between two parties after they have exchanged some random generated data.

Digital signature: A data value generated by a public key algorithm, which is based upon the contents of a block of data and a private key, yielding a individualised cipher checksum.

Down line loading: A method of key/parameter distribution to cipher machines by means of a secure channel.

Dongle: An electronic access device.

Electronic code book (ECB): A block cipher that consists of applying a cipher, or code to block of data in sequence, one block at a time.

Electromagnetic compatibility (EMC): The stray electromagnetic radiation (noise) given out by an electronic device that may adversely affect the operation of another device.

E-mail: Electronic mail protocol for sending messages between users of a network.

Encapsulating security payload (ESP): A data packet that is entirely encrypted, including the address header, to which another header is attached for the purpose of hiding the original header.

Enigma: A German cipher machine that used a series of wired rotors to encrypt messages for data transmission, during the Second World War.

Exclusive OR: A computational device, often in the form of an electronic gate, that adds two bits together, i.e. modulo 2 addition and discards any carry on.

Exhaustive key search: See 'Brute force attack'.

Firewall: A device that is installed at a point in a computer network where data flow in and out of that network and control that flow according to the rules programmed in the device.

Integrity: The ability to ensure that information has not been modified except by people who are explicitly intended to modify it.

International Data Encryption Algorithm (IDEA): A block cipher algorithm developed in Switzerland.

Internet protocol: A protocol that carries individual packets between hosts.

IP address: The host address used in IP transmission.

Key distribution centre, key management centre: A device that provides secret keys for a secure network and organises the distribution of those keys throughout the network components.

Key encryption key (KEK), key transport key (KTK): A cipher key that is used to encrypt session and/or data keys but is never used to cipher data payloads.

Key escrow: A mechanism for the storage of cipher keys, so that a third party can recover them if necessary and use them to decipher the other party's ciphertext.

Key length: The length, in binary digits of a cipher key. Typically 56, 128, or 256.

Key stream: The output of a key generator that is used to convert plaintext into cipher text and vice versa.

Key stream period: The time taken for a key stream to repeat itself.

Local area network (LAN): A network that consists a single type of data link that resides within a physically specified area.

Masquerade: A method of attack whereby an entity takes on the identity of another user without authorisation.

Message authentication code: A method of authenticating text or data messages by the use of encrypting keys.

Modulo 2 addition: The binary addition of two bits, by an exclusive OR function.

National Security Agency (NSA): An agency of the US government that is responsible for the interception of communications for intelligence reasons and for the development and control of cipher systems to protect the government of the USA.

Non-repudiation: The inability of a message signatory to deny that the message came from him/her, by the use of public key encryption.

One-time pad: A Vernam cipher in which one bit, or character, newly and randomly generated, is used for every bit, or character of data.

One-time password: A password that can only be used once.

One-way hash function: A hash function for which it is infeasible to construct two blocks of data that yield the same hash value.

Over the air (OTAR): A method of key/parameter distribution to cipher machines by means of a secure channel otherwise called 'over the air re-keying'.

PC card (PCMCIA): A standard plug-in peripheral that is often used in laptop computers and can be adapted to function as modems or as cipher modules containing algorithms and other sensitive parameters.

Pretty good privacy (PGP®): An algorithm written by Phil Zimmerman to provide a high standard of encryption for the general public, amongst others. Free versions are widely available on the Internet.

Private key: A key that is one part of a key pair, used in public key cryptography that belongs to an individual user and must be kept secret. Data ciphered by a user's private key can only be deciphered by that user's public key.

Public key: A key that is one part of a key pair, used in public cryptography that is distributed publicly. Data ciphered by a user's public key can only be deciphered by that user's private key.

Public key algorithm: An asymmetric algorithm that uses a pair of keys, a public and a private key for ciphering and deciphering.

Random number: A number whose value cannot be predicted.

Red designation: A designation given to cables, components, equipment and systems, which carry classified signals.

Red/black separation: A design concept that separates parts of a system carrying plaintext from parts that carry ciphertext.

Replay: An attack whereby an intercepted message is retransmitted with the intent of confusing the receiver of the legitimate message.

Rivest, Shamir, Adelman (RSA®): A public key system that can encrypt or decrypt data and also apply or verify a digital signature.

Router: A device that carries IP packets between networks and is used to direct those packets to the next station in the transmission route.

Secret key: A cipher key to transform a plaintext into a ciphertext and vice versa.
Server: The device in a network that provides services to clients and other entities on the network, e.g. printing services.
Session key: A cipher key that is intended to encrypt data during a limited period of time, typically for a single transmission after which the key is usually discarded.
Spoofing: Similar to masquerading, i.e. pretending to be somebody else.
Stream cipher: A cipher that operates on a continuous data stream instead of processing it block by block at a time.
Symmetrical algorithm: A cipher algorithm that uses the same key for encryption and decryption.

Tamperproofing/resistance: The technique of providing logical and physical protection to a cipher machine or module, rendering it infeasible to attack.
Tempest: The term given by the US government to identify the problem of compromising radiations.
Time authentication: A technique used by cipher machines to remove the threat of message replay.
Transmission control protocol: Internet protocol that supports remote terminal connections.
Triple DES: A cipher that applies the DES cipher three times.
Trojan horse: A program with secret functions that accesses information without the operator's knowledge and is usually used to circumvent security barriers.
Tunnel mode: ESP mode that encrypts an entire IP packet including the original header.

ULTRA: The code name used to describe a British code-breaking system during the Second World War.

Vernam cipher: Cipher developed for encrypting teletype traffic by computing the exclusive OR (modulo 2 addition) of the data bit stream and key bit stream as commonly used in stream ciphers.
Virtual private network: A secure communication system that uses encryption to exclude all other users and hosts from the 'network'.
Virus: A small program that attaches itself to a legitimate program so that when the latter is being run, the virus copies itself to another legitimate program.

Wide area network: A network that connects host computers and sites across a wide geographical area.

Acronyms and Abbreviations

AES	Advance encryption standard
ASIC	Application specific integrated circuit
ATM	Asynchronous transfer mode
AuC	Authentication centre
BS	Base station
BSC	Base station controller
BSS	Base station system
BTS	Base transceiver station
CCD	Charged couple device
CEPT	Conférence des Administrations Europèenes des Postes et Tele-communications
CMOS	Complementary metal oxide semiconducter
DES	Data encryption standard
DCE	Data connection equipment
DCN	Data communications network
DLL	Down line loading
DTE	Data terminal equipment
EMC	Electro-magnetic compatibility
FM	Frequency modulation
GSM	Global system for mobile communications
HF	High frequency (3–30 MHz)
IDEA	International data encryption algorithm
IV	Initialisation vector
INMARSAT	International Marine Satellite Organisation
LES	Land earth station
MES	Mobile earth station
MS	Mobile station
OTA	Over the air
PABX	Private automatic branch exchange
PC	Personal computer or printed circuit
PCB	Printed circuit board
PCMCIA	Personal Computer Memory Card International Association
PIN	Personal identification number
PGP	Pretty good privacy
PSTN	Public switched telephone network

TCP	Transmission control protocol
TDMA	Time division multiple access
UHF	Ultra high frequency (300–3000 MHz)
VHF	Very high frequency (30–300 MHz)
VPN	Virtual private network
WAN	Wide area network
WS	Work station

Key abbreviations

CEK	Card encryption key for ciphering chip cards
CMK	Customer master key: a source key used to generate a session key
DK	Disk encrypting key: a key used to cipher hard or floppy disks
FLK	Future link key: a source key to be used to generate a session key for a specific communications link, at a later time
KEK	Key encryption key: a general term describing a key used to protect a message encrypting key, usually during transport
KSK	Key storage key: a key used to encrypt ciphering keys stored in memory
KTK	Key transport key/key transfer key: a key used to cipher a message encrypting key during its transport
CHK	Channel or link key: source key used to generate session keys for a specific link
MCK	A key, often link specific, that is used as a source key for the generation of session keys. See CMK
MK	Management database key: a key used to cipher management data on a key management centre
NCHK	Next channel key: a link key for future use
PaCHK	Past channel key: a link key that was used in the past
PrCHK	Present channel key: a link key that is being used at present
SK	Secret key: the data encryption key
SK-B	Secret broadcast key: a data encryption key used to cipher data simultaneously to a number of stations
TRK	Tamper resistance key: a unique, logical protective key ciphering sensitive data within a tamperproof/resistant module
Xx	Default keys

1

Threats and Solutions

History tells us that, in the past, the confidentiality of data, whether it be voice or text, is the much sought after property of a communications system. Despite the great efforts made at ensuring confidentiality during the years up to, including, and in the aftermath, of Second World War, when the consequences of the interception of sensitive information were dire, confidentiality was all too readily compromised. In recent years, many documents of that era, previously hidden away in the vaults of various national security agencies, have been released for public digestion, and there is much to be learnt by security managers in studying these historic texts. Whether it be by cryptanalysis, traffic analysis, subterfuge or supposition and often by a combination of one or more, highly sensitive data were available for those who cared to search for it. That is certainly not to decry the gargantuan and inspiring efforts of both Allied and Axis communications specialists alike, far from it. It is rather an amazing fact, however, that 'so much was read by so many' during that era. So astonishing were the analytical results of the British Commonwealth countries and the USA communications code breakers that it took some time for the fruits of their efforts to be both recognised and appreciated for what they were. Eventually, it was commonly acknowledged that the code breakers shortened the Second World War by some two years and, without doubt, saved many thousands of lives.

By far the weakest link in communications was the confidentiality of Morse text messages by radio transmission, or rather the lack of it! Both allied and axis forces exhibited unbelievable naivety in the face of compelling evidence that the opposing forces were reading each other's mail. There were occasions when the integrity of messages was exploited and authentication falsified, but the most effective ploy was passive eavesdropping coupled with traffic analysis. In the battle of the Pacific, the Japanese never succeeded in breaking a major American code, and their inability to do so convinced them that their own codes were secure. Their reluctance to change codes more often than they did was largely influenced by this mislaid confidence. However, the problem was further exacerbated by the broad geographic distribution of their forces, and this led to a major failure in implementing effective key management. It was a costly weakness. Similarly, the German high command was so confident of the impregnability of their *Enigma* machines that they too were often 'casual' about their operational discipline. When key changes were made, the initiative was undoubtedly wrested away from the code breakers. Unfortunately for the Axis forces, key changes were often either infrequent or badly managed and operator proficiency at times was catastrophic. These failures severely undermined the integrity of the whole security strategy.

The cryptographic battle of Midway was won by the American analysts, with the consequence of this being the loss of three Japanese aircraft carriers to air attack. This was only made possible by the ships being located by American intercepts of Admiral Yamamoto's signals to his fleet. For many, this was the turning point of that theatre of the war, and once again, despite persuasive evidence that the Japanese communications had been compromised, little retrospective action was taken. Even the American press at the time (much to the chagrin of the military) reported on the Midway success and strongly inferred that the successful outcome of the battle was due to the compromising of Japanese ciphers.

Parallels can be drawn with the European theatre, for, in 1940–1941, the British mainland was in grave danger of being invaded and was under constant attack by the *Luftwaffe*. The German initiative was grossly undermined by the fact that the *Luftwaffe* communications security was notoriously weak, and the 'Battle of Britain' ensued, with the German airforce being defeated, despite their overwhelming numerical superiority. Even the submarine forces of Admiral Dönitz, which threatened Britain by strangulation of its sea borne supplies, were eventually subdued after being very close to success in bringing Britain to its knees. The German naval *Enigma* was almost certainly the best managed of their security networks, and it was not until considerable effort and the capture, better described as 'snatches' of two naval Enigma's from sinking German submarines by the British Navy and as portrayed in the recent Hollywood film 'U501', as a result of American Naval action, that the tide was turned in the Battle of the Atlantic. As with the Japanese, Dönitz, after suffering the sudden and startling loss of a large number of submarines and their '*Milch Cows*' supply vessels, actually questioned the confidentiality of his communications. Yet, despite the salient evidence supporting his fears, he chose to take little remedial action, but when he did respond with an extra wheel for his cipher units, the success experienced by his fleet improved dramatically. Unwarranted confidence in the *Enigma* ciphering machine was to totally undermine the German war efforts, and when the four-wheel *Enigma* was broken by the British code breakers, the U-boat threat never again seriously troubled the life-giving Atlantic convoys between America and Europe.

The Allied forces were also negligent in their efforts to maintain secure communications. Churchill himself is reported to have been less than diligent when engaged in transatlantic 'hotline' discussions with President Roosevelt on a line duly tapped by the German eaves-droppers. Even more dramatic was that despite having first-hand knowledge of code breaking successes of their own analysts, both the British and American forces were arrogant about their communications. As a result, both lost many men and especially ships by failing to appreciate frailties in their own communications. Of particular embarrassment were the losses of the British Navy, e.g. HMS Glorious, Ardent and Acasta with 1500 men, as a result of the German interception and decryption of the British Naval radio traffic and the failure to react to evidence that an attack on those ships was imminent. Even more astonishing were the losses to U-boat attacks of 347 American ships on their own 'back yard'. The most frustrating thing about these incidents was that both actions had been most accurately predicted by the 'ULTRA' organisation at the home of British GC&CS, the Government Code and Cipher school. That preventative action had not been implemented underlines the ignorance and arrogance of the ill informed and their supporting infrastructures.

So, what is to be learnt from this fascinating era of 'secure communications'? We should be aware that telecommunications are vulnerable to attack and that the threats to our communications are those to the Authentication, the Confidentiality, the Integrity and the Access to sensitive data and encryption devices. This first chapter serves to elaborate on these definitions, and the question of managing security networks and infrastructures is addressed in the final chapter of the book.

Budding security administrators and their peers and pretenders, concerned (or not!) about security, would do well to read the excellent books on this fascinating subject by David Kahn 'The Codebreakers' and Michael Smith's 'Station X,' which was screened by Channel 4 in the United Kingdom, and his subsequent publication 'The Emperor's Keys' and learn from the lessons graphically portrayed within.

When considering the security of any communications medium, there is a fundamental question to ask before any steps can be taken to analyse and implement security tools. That question is: 'What is the value of my secrets or the information that I rely upon for my comfort or existence, and what are the consequences of its loss?' There are many degrees of threat as there are of solutions to those threats, but the answer to the question is the guide to what lengths should be taken to secure the user's position. Generally speaking, there are three levels of information security, i.e. personal security, commercial security involving financial transactions and trade and high security, which encompasses national security, i.e. political and military security.

Telecommunications technology continues to advance with great pace and momentum and as the technology expands, as do the threats to those communications. There is a seemingly eternal battle taking place between those who wish to protect their communications lines and those who wish to invade them. Between the *cryptographer* and the *cryptanalyst* and between the security manager and the intruder, as each side seeks to gain the initiative over the other. The burden of ensuring the security of a network falls upon the shoulders of the security manager, and it is an onerous assignment. It is also the security manager's responsibility to ensure that all users of their network are aware of the threats and that they are instructed accordingly to follow the procedures and guidelines. Once a security policy has been decided upon, it is the task of the security manager to develop the tactics that suit their applications

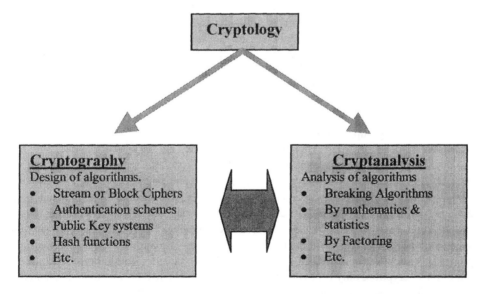

Figure 1.1 The components of cryptology

best and to implement those tactics by setting up an efficient infrastructure and formulating operational procedures that will maintain their network communications security. The threats that exist are in both human and technical forms. The latter is the subject of this chapter, with the problems involving human resources being treated in greater depth in Chapter 14. Figure 1.1 is a useful introductory guide to the competing entities within Cryptology and introduces some of the expressions that are used throughout this book.

1.1 The Technical Threats to Communications Security

Generally speaking, the threats to communications, which have existed since man started sending messages, are eavesdropping, modification, replay, masquerading, penetration and repudiation, and the means to achieve these have evolved as highly sophisticated techniques. The cryptographic countermeasures or 'security mechanisms' to meet these threats are classified as:

- Authentication
- Confidentiality
- Integrity
- Availability

1.2 Authentication

The first problem about any message being transmitted or stored, whether it is a voice or a text message, is the question of its authenticity. Is this message coming from the purported source? In voice transmissions using reasonably high-quality transceivers, voice recognition is the most obvious authentication method where the receiving party is familiar with the voice of the caller. However, where speakers are unfamiliar with each other and perhaps when the voice quality of the medium is not as it might be, other measures need to be taken to authenticate the caller and receiver. The applications discussed in Chapters 3–6 further illustrate how the problem can be largely overcome by encryption and essentially by suitable key management, as illustrated in Figure 1.2. With either a symmetrical (i.e. same keys at each end of the communications link) or asymmetrical algorithm (i.e. dissimilar key components at each end), the bases A and B can be certain that they are the genuine parties to the call as only they have the same key. If, however, the key being used to cipher the call (voice or fax) is common to a group within the network, then one can only be certain that the calling parties belong to the same group. This may well be sufficient for the network in question, but it is for the network security manager to consider this when organising their key distribution.

There is a loophole, however, which may be exploited and that is the ploy of 'replay' or 'spoofing', whereby a third party taps into the link, records the transmitted message and then retransmits it at a later date. Unless the eavesdropper has the correct security equipment and the proper keys, they will not be able to read the message. However, the retransmitted message will introduce confusion at the intended receiving destination. Consider the example in Figure 1.3, where station A transmits a voice message 'attack' to Station B at 9.00 a.m. As a result of the encryption, only B having the corresponding key will be able to understand the message. Station Z, the eavesdropper, will not be able to understand the message but will be

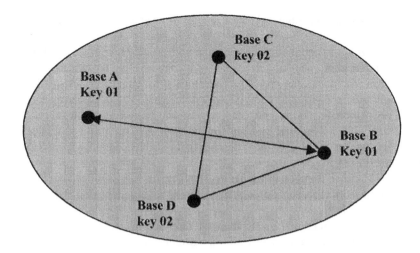

Figure 1.2 Message authentication by possession of common keys

able record it. If Z then retransmits the message 'attack' at 3.00 p.m., one can imagine the ensuing chaos caused at station B by the reception of what appears to be an authentic message. It is, after all, encrypted by the correct secret key. To overcome this method of attack, time authentication must be included in the security package, and when implemented, Station B will not receive the 'replayed' message as the cipher at B unit will not be able to

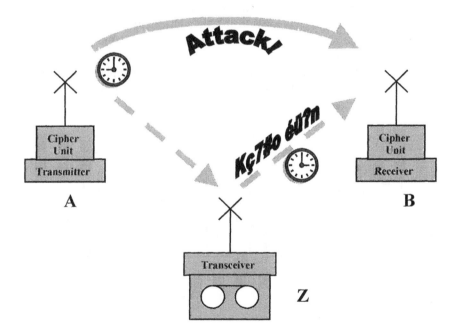

Figure 1.3 The need for time authentication

synchronise with the late message and hence will never be able to read the later version or the message.

Time authentication is one method of message authentication and is often found in voice and fax encryption equipment and is certainly a tool to look for when considering the purchase of such machines. The protection is achieved by either introducing a time slot of typically 5 min after the original encryption, within which the deciphering machine must perform the decryption, or modifying the key generator process so that the generator at B will not synchronise with the original generator position at A. The 5-min time slot is usually sufficient to account for any slight time difference in the machine settings around the network. In other words, all machines in that network must have the same time ±5 min. The use of time slots is trickier than it might at first seem. The receiver station must have the capacity to check several time slots at the same time as two stations having very similar times can, in fact, be in different time slots.

Other authentication methods exist, such as time stamps and mutual key agreement mechanisms, and each has their niche within a particular message system.

1.2.1 Text/Data Message Authentication

As most text or data messages are not 'real time' communications, a different method of authentication is required. This is known as the message authentication code (MAC), and the process is illustrated in Figure 1.4.

Authentication by encryption with symmetrical keys has its limitations, as inferred above. However, the application of asymmetrical encryption using the RSA® (Rivest/Shamir/Adleman) algorithm guarantees the authenticity of a message by the fact that, as described in Chapter 2, the asymmetric algorithms are founded on two key-pair components: one, the *private* part of the key and two, the *public* part. The authenticity of the message source is guaranteed because if the ciphered document can be deciphered by the *public key*, it must have been ciphered by the partner's *private key*. This follows, as only the original owner has possession of the *private key*, and the message can only have been ciphered by that person.- Conversely, a message ciphered by the *public key* can be deciphered only by the owner of the

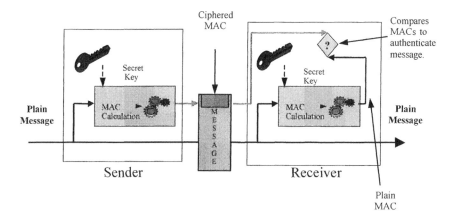

Figure 1.4   The MAC process

corresponding *private key*. However, this latter case highlights a flaw as far as authenticity is concerned: *any* possessor of a public key can encrypt a message for the owner of the corresponding private key, and therefore, the source of that message is not certain.

The authentication process can be achieved by the use of a MAC with either symmetrical or asymmetrical keys, as indicated in Figure 1.4. The MAC is similar to a hash function except that a virus can be used to modify a hash function. The MAC, however, cannot be modified in the same way as it relies on a key known only to the users. The secret key ciphers the MAC, attaches the result to the message and forwards it to the transmitter. On reception, the encrypted MAC header is removed from the message, deciphered by the secret key and the resulting calculation compared with the original plain MAC value from the message to check the message's integrity.

1.3 Confidentiality

The confidentiality of a message, voice, text or data is assured by encryption with a secret key, provided that only the legitimate users have access to that key. Symmetric encryption, therefore, can provide confidentiality of a message. An eavesdropper might well have access to the cipher text, but unless they are in possession of the correct copy of the encrypting key, they will have no opportunity to read the plain text. As we shall see later, the secret key (SK) in a symmetric system is common to both sender and receiver. An asymmetrical algorithm may also be used to carry out the encryption, but in this case, the keys are not common to both parties. There are, however, strong arguments for symmetrical keys rather than asymmetric keys being used for the purpose of encryption for confidentiality, the main reason being that symmetrical encryption is faster than asymmetrical. However, as the characteristics of both methods of encryption are useful in message protection, hybrid systems are very often adopted, combining their advantages (see Figure 1.5).

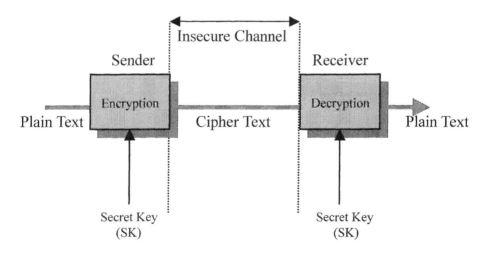

Figure 1.5   Confidentiality by symmetric encryption

1.4 Integrity

Messages and files need to be protected against surreptitious modification, and whilst confidentiality procedures protect against eavesdroppers, they give little protection against modification and the *integrity* of the message or file. This is critical for text and data messages, which are vulnerable to this form of attack. This is especially the case in the banking and other financial arenas where an intruder may be able to change monetary values and account numbers in a standard, transaction form without needing to actually read it. The solution to integrity threats is to employ digital signatures, MACs or some other redundancy scheme in the plain text and then use encryption.

1.4.1 Digital Signatures

These are asymmetric encryption tools that allow the author of the original message to 'sign' their document in such a manner that the receiver can verify that what they receive is a faithful copy of the author's original. The procedure is illustrated in Figure 1.6. Any modification of the protected message in transmit will result in the derived signature being different to that of the original, proving loss of *integrity*.

Using the RSA® system, the sender signs their plain message with their *private key* and

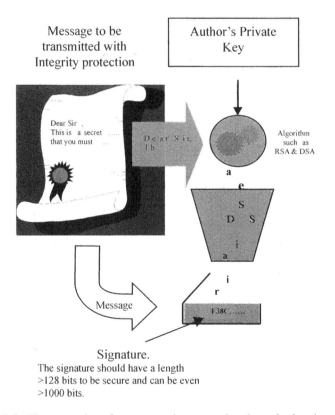

Signature.
The signature should have a length
>128 bits to be secure and can be even
>1000 bits.

Figure 1.6   The generation of a message signature using the author's private key

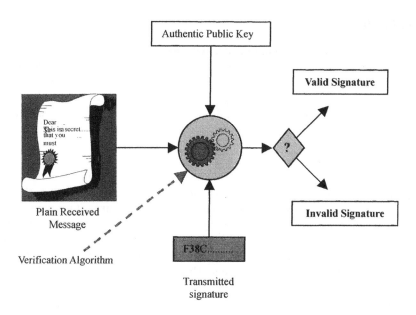

Figure 1.7   The receiver runs a verification algorithm to detect modifications to the message

transmits it, along with the message, to the receiver. The receiver being in possession of an authentic copy of the *public key* of the *key pair* is able to compare the original signature from the sender's document with that of the received message. This is done by running the verification algorithm with inputs of the authentic public key, the plain message and the sender's original signature. If, during the transmission through an unsecured medium or channel, the message has been tampered with, the verification performed by the receiver will give the output 'Invalid Signature' (Figure 1.7).

The purpose of the digital signature is just to check the message integrity. It is not used to encrypt the message and therefore does not offer confidentiality. However, combining the two techniques, where symmetrical encryption of the message text ensures confidentiality and with signature verification, by public key techniques ensuring message integrity, a hybrid system is produced. The result is a very powerful tool in protecting files and messages.

Furthermore, the use of public key encryption to generate and verify the signatures imparts authenticity on the message as only the possessor of the *private key* could have signed the original text, if their *public key* verifies it. Conversely, the originator having signed with their *private key* cannot deny having done this, as only they are in possession of their private key. This imparts the feature of non-repudiation

In summary, then, digital signatures offer:

- Public verifiability: where anybody in possession of the authentic public key can verify the signature
- Authenticity and integrity: as modification of a message or replacement can be detected
- Non-repudiation: the signatory of a message cannot deny having signed the document

There is a further discussion on the subject of digital signatures as asymmetric algorithms in Chapter 2 and also in the applications modelled in Chapters 10–12.

1.5 Availability

One of the more basic, yet essential, fundamentals in communications security is the control of availability and of access to the medium, sensitive data and ciphering equipment. The subject of physical access to the premises containing these entities is certainly an important issue, but as this book is focused on cryptographic security, physical access to buildings, etc. is beyond its scope. However, there is some discussion of physical access to security modules here and throughout the application chapters.

1.5.1 PINs and Passwords

The purpose of a password and PIN system is to authenticate users and facilitate their right of entry to whatever functions they are permitted to employ. In principle, it is a simple and basic method of controlling access, yet it is surprising that these tools, which form an essential part of the security process, are the subject of much apathy and abuse. Throughout the author's experience in security projects, he has been constantly amazed at the naïve discipline and application of this basic security instrument. However, when considering the number of passwords that an individual must remember, it is perhaps not so surprising that they might select passwords that are easy to remember. Once having become familiar with a password, people are reluctant to change them. Today's businessman or businesswoman is required to remember:

- Mobile phone access
- Mobile phone lock
- Mobile phone provider customer's password
- Personal e-mail password and username (possibly numerous?)
- Bank ATM cash withdrawal password
- Credit-card PINs and passwords (numerous?)
- Office-door password
- Company e-mail password and username
- Company LAN access password and username
- Favourite user-group WEB page passwords
- Briefcase lock combination
- Data security passwords and pass-phrases when using tools such as Pretty Good Privacy (PGP®)
- E-banking contract and password numbers

So, it is small wonder that the normal executive often has password problems. Add to these the supplementary passwords that a security operative might be expected to be familiar with, and we can easily see the factors that invite over simplification and lassitude. Although it is in one's interest to guard personal data carefully, there is a tendency, when confronted with this mass of alphanumeric data, to relieve the situation as far as memory capacity is concerned, by resorting to either of the following:

- Writing all passwords in a diary
- Assuming the same password for all applications
- Relating the password to the particular application, e.g. using the floor and room number as the access to the office door.

- Using very simple configurations such as: 11111111 or 12345678, etc.
- The dangers are mostly obvious.

Diaries can be easily lost and are an obvious target for anyone seeking to gain compromising personal information of an individual. It is far better for those travellers with a laptop to make a file to contain all the passwords and then protect that file by encrypting it with a symmetrical PGP® key generated from a single pass phrase. Hence, 'one protects all', but of course, all becomes vulnerable by an attack on the master password or pass phrase. For those who are compelled, for one reason or another, to write things down, they should at least make life difficult for the trespasser by juggling the characters, listing them in reverse order or subtracting each digit of a pass number from 10 and logging the result instead of the password. However, these are trivial precautions and should not be used where high security is required. Any worthwhile security administrator would be horrified by this rudimentary action.

Most people at some time or another have been familiar with the 10-digit door pad, security lock. On first consideration, it might seem a formidable task for the uninitiated to 'break the code' and gain access to the treasure hidden inside. For the more flexibly minded, though, it is far from a considerable task. A *brute-force attack* or, as it is otherwise known, an *exhaustive key search,* is faced with trying all possible numbers until the correct sequence is found. Normally, a four-digit password is used, and this gives a *key variety* of 10,000, i.e. 10^4 when the digits 0–9 are used. With no delays inserted into the brute-force attack on the door pad, any individual can cover all possibilities in, say, about 14 h. So, a weekend guard or cleaner would have no problem attacking the pad successfully. Of course he/she could get lucky, find the solution within the first few attempts and so prove, to some extent, that the brute-force statistic is misleading. A cryptanalyst tackling the door pad access would look for alternative solutions, especially when confronted with a large *key variety.* In the door-pad model, they would look for clues that might offer more profitable dividends. Checking on the door or floor number, the occupier's birthday, their spouse's birthday, telephone numbers and car registration plates are all prime possibilities for access codes. An inspection of the condition of the pad itself, e.g. dirty fingerprints, which would probably identify the four digits used, if not the order, presents a different method of attack. Permanent passwords or PINs would leave worn digit buttons, and a host of other clues make the assailants task that bit easier. Combining these alternatives generated by lateral thinking can reduce drastically the time to carry out a brute-force attack. So, the code breakers of Bletchley Park, the home of British code breaking during the Second World War, and even those of present eras, looked for a toehold, a chink in the armour to make inroads into security parameters, and it is left to the vigilant security manager to make life as difficult as possible for those wishing to gain the secrets of their charges.

Extrapolating from the model above, PIN and password access is best controlled by adopting a policy of central command, whereby the central body controls all passwords and PINs from their generation, through use, enforced changes and eventual destruction. This tactic is far more secure than relying on individual network personnel using their own judgement of what and for how long a password should be used. Once an individual learns to remember their PIN, they are reluctant to change it, and the repetition of sensitive data such as passwords or parameters represents a gift to the assailant. A prime example of this led to the daily rotor settings (daily key initialisation) of the Enigma machine becoming predictable to the analysts. Left to the individual users to arrange, the daily start codes provided a toehold into breaking the system. Such gifts can be considered as the pieces of a jigsaw or crossword puzzle, and eventually, when enough evidence has been gathered, the overall picture is there for all to see.

1.5.1.1 Guidelines for Password Use

Passwords should be centrally controlled wherever possible but, in any case, should follow the guidelines below in order to add strength to access security:
Passwords:

- Should be kept absolutely secret and not divulged to any other user
- Should not be written down or recorded where they can be accessed by other users
- Must be changed if there is the slightest indication or suspicion that a password has been compromised
- Must be changed when a member of the organisation leaves the group or changes their task
- Should use a minimum of six alphanumeric characters
- Should not be formed from any obvious source, e.g.

  - Username or group/company/project name
  - Family name or initials, or partner's name
  - Months of the year, days of the week
  - Car number plate registration
  - Nicknames/pet names
  - Telephone numbers
  - All numeric or all alphabetic characters
  - More than two consecutive identical characters

- Must be changed monthly or at least bi-monthly.
- Must be changed more frequently the greater the risk or more sensitive the assets they protect
- Must not be included in an automated log in procedure, i.e. not stored in a macro function

1.5.1.2 Guidelines for Password Management

Password management systems should provide an effective, interactive resource that ensures the quality of the passwords and enforces their use according to the security manager's policy. Generally speaking, password management should enable secure login procedures and protect passwords from unauthorised use and access. This includes precautions taken to ensure that passwords are stored in files that are separate from main application system data and that they should be stored in an encrypted form, by a one-way encryption algorithm. This is an algorithm that takes an input string and encrypts it at the output. This is a relatively easy process, but the reverse operation is intended to be infeasible, or at least very difficult. These measures offer some protection against 'password cracker' programs or 'dictionary attacks'. The dictionary attack seeks to carry out a brute-force attack on an encrypted password file by comparing the file contents with a pre-defined list of simple passwords (usually of many thousands), which are also encrypted by a one-way function to find a match. In practice, dictionary attacks have the reputation of having some success, and so password files must be still considered as being vulnerable.

Initial or default passwords from manufacturers must be replaced after equipment installation and form part of the separation process between the client and the producer. There should be no access to a protected system without the correct password submission, which must be enforced by the management. There are cases for the individual user to be able to select their own passwords, and, in such cases, a re-confirmation, by retyping a new password definition, should be made compulsory. Password changes should be enforced at predetermined intervals and a record of them kept so that they may not be recycled.

It is apparent that whilst users may be given the choice of password data, password policy and implementation should be centrally controlled and formally managed by the following process:

- Users should sign a declaration, undertaking to keep personal passwords confidential
- Passwords should be conveyed in a secure manner and therefore should avoid distribution by:

  - Telephone
  - Third parties
  - E-mail
  - Normal internal mail
  - Users should acknowledge the receipt of passwords

- Initial passwords should be forcibly changed after their first use
- Temporary passwords should be issued on the occasion of a user forgetting their password

Apart from the access discipline of security personnel, there are numerous logical approaches to gain access to security equipment and protected data. The *challenge/response* procedure is an accepted method of dealing with the problem.

1.5.2 Biometric Access Tools

The advance of biometric tools as a means of personal identity, as per James Bond epics, is now not so far fetched as it was a few years ago. The main areas of interest lay in the study of:

- Iris and retinal identification
- Vein patterns
- Fingerprints
- Speech recognition

Increased computer processing power, coupled with advances in electronic sampling and scanning techniques, has made all of the above physical features ideal for personal recognition. These characteristics are as unique as an individual's genetic code, and as such, each of them has become an excellent tool of personnel authentication. Of course it is their genetic code that is the source of these characteristics. The applications of biometric identification conjure up images of keyless cars, biocredit cards and a world free of the haunting password dilemma. However, to the general public, which might be pondering the question about why we still see little evidence of biometric ID application? Whilst the idea appears to give an ideal solution to communications access, biometrics has some way to go before the days of the PIN and password become a distant memory. At the time of writing, cost efficiency and lingering techno-hitches largely leave passwords as the most reliable identifying tool for the immediate future. However, fingerprint and iris scanning seem to be the best biometric bets as future, personnel authentication mediums as far as security is concerned.

Fingerprint IDs can actually be ascertained in a number of ways as they exhibit a layered structure, each of which can be examined by different scanning techniques. The outermost layer carries the familiar shapes, the arches, loops and whorls, etc. The second layer supplies more unique features, i.e. the ridge structure and the bifurcate divisions. By taking into account the position, direction and orientation of these characteristics, a complete identifying package can be constructed. The less obvious third layer is defined by the pore structure; this too can be scanned and used by data banks to compare the scanned digit with the images stored within a computer's data banks. The fingerprinting tools include the traditional inkpad and paper, though for communications access and authentication processes, this method is not practical. Electronic scanning mechanisms are much more to the point. Charged coupled devices (CCDs) and CMOS, semiconductor chip, scanning arrays that can be physically implemented into cipher units, lend themselves to device access monitors. There is a general consensus amongst informed technocrats that the automatic scanning of a fingerprint or iris pattern will be the means of logging into our personal data sanctuaries, but the use of biometrics as high-security access tools is still a cause for concern. It is an exciting idea that our personal unique physical qualities can be used in secure communications, not just as the 'ultimate' access tokens, but also perhaps as actual ciphering keys. Imagine, for the moment, being able to pick up a telephone or radio handset and cipher the call by virtue of the fingerprint of the hand that holds the transmitter. For symmetrical encryption, as fingerprints remain with us for the duration of our lives on this planet, then there is limited scope as ciphering keys in high-security systems as we would always be using the same key. At best, those blessed with a full complement of digits would only be able to rotate keys around a 10-day cycle. Perhaps, a more realistic application of encryption by biometrics would be found in asymmetrical encryption whereby a fingerprint forms a seed for a private key within an asymmetric key pair, or in key agreement. The idea brings a whole new meaning to digital encryption. However, before embarking upon a mission to discover a new method of securing data, there is a fundamental problem with human biometric scanning as a security contrivance. It is that fingerprints of an individual are easily obtained, surreptitiously or otherwise and can just as easily be copied. The same criticism is true for voice, vein and iris/retinal scans. Perhaps genetic identity is the next avenue for exploration? For the present day, however, biometric-based encryption remains in the '007' realm, but as access tools, biometrics have already found a useful niche.

1.5.3 Challenge/Response Control

This form of access control seeks to resist the threats to user authentication by such activities as spoofing, i.e. an impostor pretending to be the legitimate user. The system is based upon *something known* to the user (password or PIN, etc.) and *something possessed* (a chip card, dongle or the like).

Consider Figures 1.8 and 1.9. The user commences their entry procedure, which might be inserting a smart card into an encryption device or a remote computer function to access files. The destination unit generates a true random number, which is transmitted to the user's

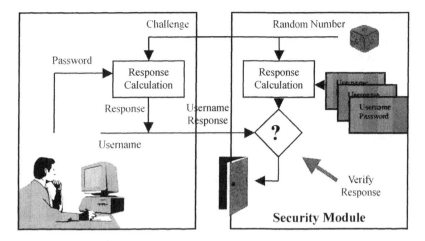

Figure 1.8   The challenge/response method of access control

terminal as a 'challenge'. The user then enters their password, and these two values are presented to a cryptographic algorithm, e.g. a hash function that generates a response result to the inputs of the challenge and password. The resulting 'username' response is transmitted back to the source security module where the remote username response is verified by comparing it with the expected value stored in the source module. Upon successful corroboration, the user is allowed access to the desired function.

1.5.4 Tamperproof Modules

Physical access to security modules within their host devices can render the sensitive contents vulnerable to attack and monitoring whilst in an untrusted environment. The threats to be guarded against are:

- The readout of cryptographic data such as keys and implementation of algorithms.
- The modification of cryptographic data in order to influence the encryption process in a manner beneficial to the invading party.
- The modification of cryptographic data in order to weaken the security processes.
- The input of tools such as 'Trojan Horses' to weaken the security processes.

With state-of-the-art encryption, all processes and the cryptographic data that they use are built into the security module in a permanent manner, the principle being that no cryptographic data or keys ever leave the tamperproof module in a plain condition and that they run entirely within the module. Any data leaving the module perimeter should be ciphered by a resident key, unique to that module and therefore removing the threat of read out and modification. Ideally, no copy of the resident cipher key should be made.

There are several approaches to tamperproofing equipment that vary from these examples mentioned here to those extremes where the simple movement of a device can be detected internally and action taken within the security module to render its contents unreadable. Similarly, sensors within the tamperproofing material can detect entry attempts and also take preventative action.

1.6 Compromising Emanation/Tempest Threats

Another form of access to cryptographic processes is found in the radiation emitted by all communications equipment and its security attachments, the compromising emanations.

1.6.1 Compromising Emanation Definitions

Black designation: A designation applied to cables, components, equipment and systems, which handle only unclassified signals and to areas in which no classified signals occur. Compromising emanation: Unintentional signals bearing data related, or intelligence revealing information, which, if intercepted and analysed, disclose the classified information transmitted, received, handled or otherwise processed by any information-processing equipment.

Equipment Tempest radiation zone: A zone established as a result of determined or known Tempest equipment radiation characteristics. The zone includes all space within which a successful intercept of compromising emanations is considered possible.

Red designation: A designation applied to cables, components, equipment and systems, which handle classified signals, and to areas in which classified signals occur.

Red/Black concept: The concept that electrical and electronic circuits, which handle classified, plain information, be separated from those that handle encrypted, classified information. Under this concept, Red and Black terminology is used to clarify special criteria relating to, and differentiating among, such circuits and the areas in which they are contained.

1.6.2 Compromising Emanation

Compromising emanation is due to either direct cross-coupling within the signal pass-band, caused by galvanic, capacitive and inductive coupling, or secondary cross-coupling, especially the modulation of harmonics. The latter instance is especially apparent when information signals modulate quartz-stabilised clock frequencies or free oscillating switching regulator frequencies.

Direct cross-coupling, i.e. Red/Black cross-coupling, is divided into analogue and digital signal cross-talk. With secondary cross-talk, we can distinguish between frequency, amplitude and phase-modulated harmonics. A prime example of Red/Black cross-coupling is illustrated in Figure 1.9. The cross-coupling ratio is determined relative to the amplitude of the ciphered signal. The ratio of Red analogue signals is usually measured continuously over a specified frequency range whilst the ratio of Red digital signals is determined at specific test frequencies.

1.6.3 Modulated Harmonics

Harmonics developed by the presence of clock signals are readily detected by monitoring stations. Here, as in Figure 1.10, a clock signal has been modulated by a plain, possible classified signal. There are many electronic devices that can inadvertently perform this modulation, e.g. diodes and transistors. The result is that the high-frequency clock signal behaves as a carrier for the data signal and is easily detected and analysed.

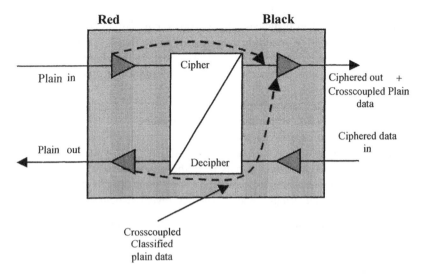

Figure 1.9 Red/Black cross-coupling

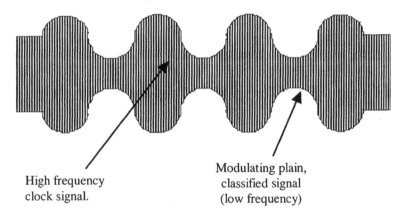

Figure 1.10 Modulation of a clock signal by a compromising plain data signal

In any secure telecommunications operation, the original message is usually in plain form, and any electronic device involved in the generation or storage of this information is liable to radiate it by some means. When this occurs, any outside party, with readily available technology, is able to monitor the original signal before it is encrypted for transmission. A common example of compromising radiation is to be found in the PC environment in which computers and their accessories are notoriously leaky. The most obvious components contributing to this open 'broadcast' of data are:

- The video monitor, which radiates the video signal generated by the electron beam as it excites the screen particles coating the internal coating materials. This is sometimes referred to as the 'van Eck' radiation, after the Dutch scientist Wim van Eck, who was the first to show that a screen display could be re-constructed remotely by the use of simple technology. There are even instances where viruses, e.g. Trojan Horses that, when operated on a PC, are capable of encoding data in such a way that it can be transmitted within the machine's Tempest emanations, i.e. 'soft Tempest' to the observer.

- The keyboard strokes, which generate signals corresponding to the plain data of a text message.
- Reading from and writing to computer hard disks.
- The computer's main units, which use graphics adaptors.
- Accessory devices such as printers and scanners.
- Building devices and materials such as heating radiators and window frames that act as antennas and propagate electronic signals from the computer.
- Power supply and telephone cables, which can carry plain signals due to 'cross-talk.'
- Power consumption of electrical devices that may reveal critical signals.

All that is required to monitor a PC workstation from a neighbouring room, building, or indeed a motor vehicle parked in the street, are some form of antenna, an amplifier, a video monitor and two time-base generators to synchronise the monitor scanning.

The term, 'tempest' was coined by the US government to identify the problem of compromising radiations. It is used as a standard of protection against electronic hardware electromagnetic radiation. The countermeasures to overcome Tempest threats are to design and construct security equipment with the following policy:

- Strict Red/Black separation, i.e. the separation of zones with sensitive data and those without. The colour coding is such that Red environments would carry sensitive data, whilst Black would carry normal 'public' information.
- Careful layout of power supplies and efficient earthing.
- Careful circuit design.
- Comprehensive filtering of all interface signals and power cables.
- Filtering of VDU monitors.
- Use of optical cables to connect accessories.
- Shielding and encapsulation of critical components and security modules.

In order to be able to manufacture Tempest and electro-magnetic compatibility (EMC)-protected communications equipment, some considerable expense must be invested in the construction of a radiation test laboratory comprising an anechoic chamber with sophisticated test and analysis equipment. The results of carrying out testing of this nature should be evident in the manufacturer's data sheet, and the failure to publish the standards normally required might well mean that the equipment does not conform to the standards such as MIL-STD-461B/MIL-STD-810E and 3G 901A that are typically required by discerning buyers. Needless to say, the cost of development and testing of computer and telecommunications equipment is eventually borne by the customers. It is a price that is well justified by government bodies who have little choice but to employ the extensive metallic shielding of individual devices and buildings as well as taking advantage of certain software techniques to plug this weak point in their security. Failure to address this problem will invite attack from interested parties who constantly search for a weak link in a system.

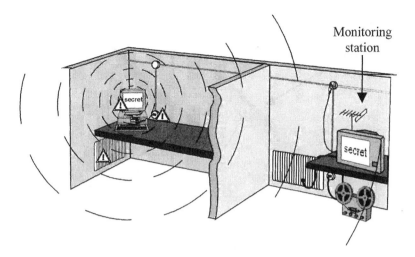

Figure 1.11 Sources and monitoring of compromising radiation

1.6.4 Electronic Coupling

Three types of coupling are briefly discussed to illustrate the problems caused by coupling:

1.6.4.1 Capacitive Coupling

This type of coupling is due to the interaction of electric fields between circuits and is shown in Figures 1.11–1.13, where two signals, perhaps on neighbouring conductors on a printed circuit board, are coupled together. The digital *Signal V1* is coupled to the analogue *Signal V2* and causes a current *ic2* to flow through the resistor *R2* and hence induce a voltage across it *Vn*. This voltage is a noise signal, but it carries the digital information of *Signal 1* and is therefore available for those detecting and analysing the second loop. The capacitor C, whilst being the coupling agent between the two circuits, may well be merely the result of stray capacitance rather than being an actual component in the circuit.

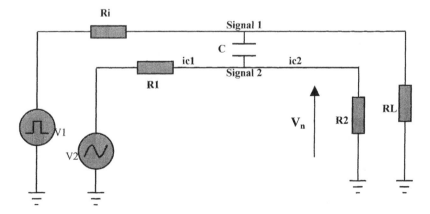

Figure 1.12 Capacitive coupling between two signal-carrying conductors

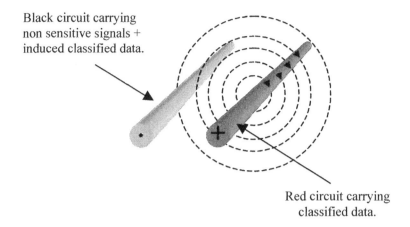

Black circuit carrying
non sensitive signals +
induced classified data.

Red circuit carrying
classified data.

Figure 1.13   Inductive signal coupling

Exactly the same result occurs when two conductors or components experience magnetic coupling, or mutual inductance. This is shown in Figure 1.13 and occurs when a current flowing in a conductor produces a magnetic field that links with the conductor or component of another circuit, thereby inducing a voltage in the second circuit. The induced voltage carries information about the data flowing through the first conductor, i.e. a *Red circuit* and is then available for monitoring and analysis if the second circuit is a *Black circuit*.

1.6.4.2 Galvanic Coupling

This occurs when the currents of two devices flow through a common impedance, i.e. Z. The current $i1$ flows through the impedance Z and the decoupling capacitor C and hence develops a 'noise voltage' Vn across the impedance Z. The Vn couples with the second current $i2$ and

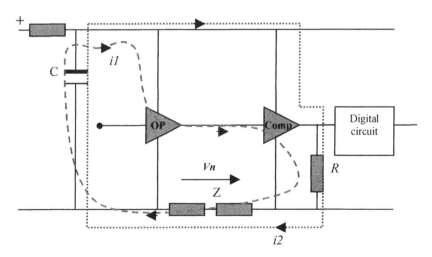

Figure 1.14   The incidence of galvanic coupling

imparts what might be sensitive data in *i1*, making it available for any analysis of that part of the circuit. See Figure 1.14. Galvanic coupling represents the most severe threat to Tempest integrity.

1.6.5 Preventative Measures in Electronic Equipment Construction

Tempest protection means that the electromagnetic energy emanating from a circuit or component will not carry any form of compromising information. In taking steps to minimise compromising radiation, the first initiative is to design carefully the layout and structure of electronic printed circuit boards, and the most effective measure is to fabricate multilayer circuit boards, as illustrated in Figure 1.15. Of particular importance in the multilayer printed circuit board (PCB) is the *null volts plane*, which is one entire layer comprising a thin metallic shield to isolate the components of the top layer from the connecting conductor and power-supply tracks mounted on the lower layers. Earthing of the power and data cables is a critical issue, particularly how and where the cable earths are linked to the null volts plane. It is also desirable that Red components be mounted as remote as possible from Black devices and circuit tracks.

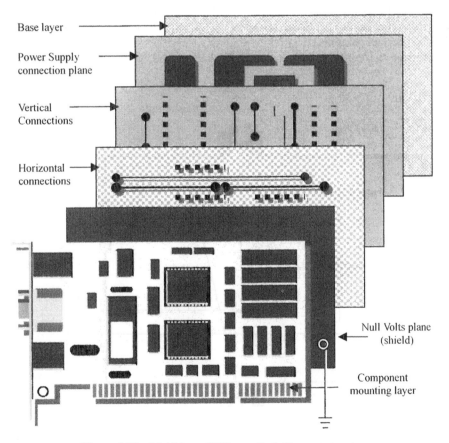

Base layer

Power Supply
connection plane

Vertical
Connections

Horizontal
connections

Null Volts plane
(shield)

Component
mounting layer

Figure 1.15   Multi-layer PCBs can limit Tempest threats

1.6.5.1 Interfacing Red/Black Circuits

Sooner or later, many Red circuits must be linked to Black devices and special care must be exercised in how the interfacing should be carried out. Fibre-optic couplers are an ideal solution, even though optical technology is not without its own problems, and even fibre-optic cables, as may be used to link PCs and their printers, are prone to leakage and corruption from external light sources.

A floating input/output interface is another solution. Interfaces such as the RS 232 interface are referenced to a common ground between I/O signals and can be the root of a Tempest hazard. Equipment earthing must be carried out in such a way as to be compatible with Tempest requirements within a system interface, and the general recommendation is to separate the *Red 0 Volt* return from the *Black 0 Volt* position.

1.6.5.2 Cabling Problems

Wherever possible, separate cables should be used for Red and Black circuitry, but there are occasions when connecting cipher units to their transceivers is not feasible. In such circumstances, it is essential that earthing impedance is kept to an absolute minimum, and this is best accommodated for by connecting shields and 0 volt points to the device ground at both ends of the cable. The length of earthing cables is also of concern, and the offending leads should be kept as short as possible.

An alternative solution is to balance the Red I/O signal, grounding it at the interfacing

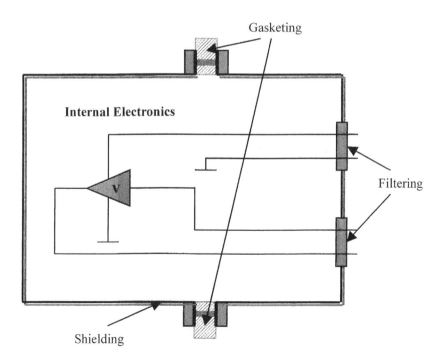

Figure 1.16   EMC protection measures in device design

device yet, once again, a common ground impedance shared with a Black circuit if the grounding of that device is not controllable.

1.6.5.3 EMC Protective Measures

Tempest and EMC problems share many features, but solving one does not necessarily mean solving the other. However, the three main steps to eliminate EMC leakage are common to Tempest. They are illustrated in Figure 1.16 and are listed as:

- Filtering of signal and power leads
- Shielding radiating devices
- Gasketing to ensure the conductive continuity of shields.

2

An Introduction to Encryption and Security Management

There are many fine publications on the subject of encryption and particularly on the features and applications of the multitude of algorithms available for that encryption. This chapter and indeed this book are not intended to compete with those texts, some of which are included in the list of references or recommended reading. However, what this chapter does offer is a brief, and hopefully a readily assimilated, presentation for the security manager who requires foundation material in order to gain a greater insight into the practical applications found in later chapters of this book. A more in-depth study of this subject is readily available elsewhere.

2.1 Analogue Scrambling

The term *scrambling* has been, and still is, used on occasions to describe the encryption process to protect voice communications whether achieved by digital or analogue means (see Figure 2.1). Indeed, some vendors of digital ciphering equipment refer to their 'scramblers' in their product technical documentation. This author's opinion, which is shared by most people in the business of communications security, is that scrambling is simply and purely a refer-

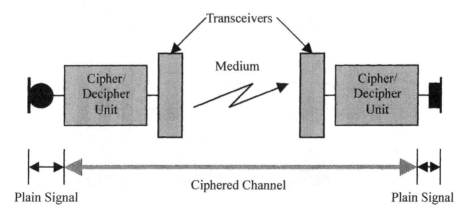

Figure 2.1 The basic features of a ciphered voice link

ence to analogue ciphering and that any digital device being marketed as such is being grossly understated as to its qualities. As far as this book is concerned, a scrambler is an analogue device.

Analogue encryption or 'scrambling' has been around for many years, and the term scrambling has become synonymous with the 'Red' telephone of 'M' in the early James Bond stories and also in the Second World War era with Winston Churchill's secure phone link across the Atlantic to Mr Roosevelt in the White House. No patriot could possibly question the confidentiality of James Bond's secret communications with 'M', but there is very little doubt that Winston's conversations with the US president were readily digested by the Axis forces through line tapping of the Atlantic cable. Sceptics are once again recommended to read Kahn's excellent record of the Second World War cryptographic battles in his 'The Codebreakers'. Here, there was strong evidence that analogue encryption in the 1940s and 1950s was extremely vulnerable to eavesdropping and analysis. Despite the great strides in the technologies of electronics and telecommunications, analogue scrambling is still today regarded with suspicion. A look at the scrambling techniques will serve to explain why this is the case.

2.1.1 Phonemes and the Structure of Voice Signals

Figure 2.2 shows a spectrogram of the words, *one*, *two* and *three*. The most prominent and indeed important features of the spectrum are the *formants*, represented by the dark regions of the graph. It is not difficult to see that the phonemes could quite easily be reconstructed from their component parts, including the formants. Therefore, if a scrambler is to be efficient as an encryption device, these frequency components must be heavily disguised, and herein lies the root of the problem. No matter what method of frequency scrambling is adopted, some voice residue remains on the transmission medium. The trouble is that each component of the plain voice signal is converted into an equal power component in the ciphered signal, i.e. the characteristic of the power function is maintained. Hence, the voice rhythm and distribution

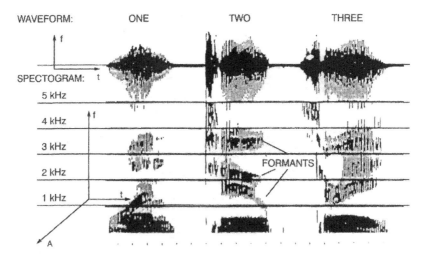

Figure 2.2 The voice spectrogram

of words and phrases can be detected in all pure frequency ciphered messages and therefore tell-tale residual intelligibility.

In order to be able to cipher a voice signal, the correlation between the three parameters of time, frequency and amplitude must therefore be altered. Where one of these parameters is changed, we refer to it as being a single dimension encryption, whereas if two or more are altered, it is referred to as multidimensional ciphering. The single dimensions are:

- Altering the amplitude axis
- Altering the frequency axis
- Altering the time axis

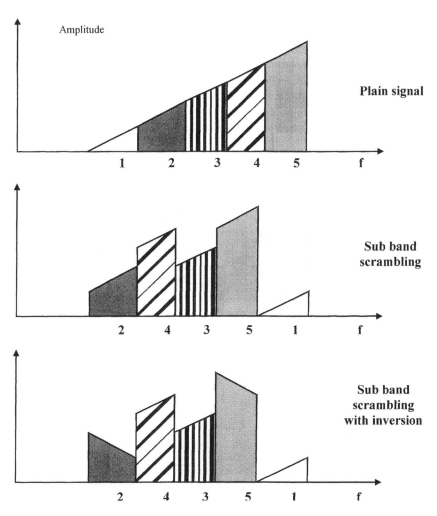

Figure 2.3   Frequency sub-band scrambling with inversion

2.1.2 Frequency Scrambling

There are several variations of frequency voice scrambling, perhaps the most common being the 'frequency sub-band scrambling with inversion' (see Figure 2.3). This is the traditional frequency scrambling method where by the plain voice message at the output of a microphone is divided into a number of sub-bands, which are then scrambled and, as an extra security measure, inverted under the control of a secret key that precisely defines the scrambling and inverting sequence. A receiving station with the same secret key available will be able to reassemble the scrambled packets of the signal and reverse the frequency inversions to reconstruct the transmitted signal.

From the five sub-bands of Figure 2.3, the total possible variations possible are:

Number of variations $=M!^*2^M= 3840$

where $M = 5$.

In cryptographic terms, this is a low number of alternatives.

The result of sub-band scrambling produces a ciphered signal that still contains residual voice patterns, which, to the eavesdropper, resemble a conversation akin to that of certain Disney characters. Although impossible to be recognised by the human ear as intelligible speech, it would be vulnerable to analysis by experts with sophisticated spectrum analysers at hand. The implementation of this form of frequency scrambling is modelled in Figures 2.4 and 2.5 and requires a *band splitter* consisting of a series of narrow band-pass filters with an scrambling switch under control of the encrypting key.

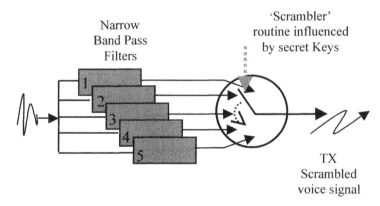

Figure 2.4   Band-splitting in the scrambling process

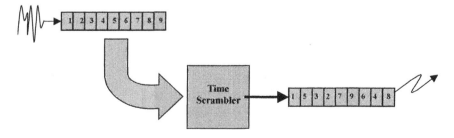

Figure 2.5   Principle of a time scrambler

2.1.3 Time Element Scrambling

Whereas the frequency scrambling above affects the individual phonemes, time ciphering influences the power distribution, voice rhythm and pause distribution of the original plain signal, whilst leaving the phonetics unaffected. It is clear that as the number of time elements used to sample the voice signal increases, the security of the method is improved. However, time scrambling introduces a delay in the ciphered signal, and this delay increases as the sample elements increase.

Figure 2.6 illustrates the principle for a combined frequency and time element scrambler that uses analogue-to-digital conversion and RAM technology to implement the scrambling. In this case, the complete signal is band split into four sub-bands of the analogue signal (see Figure 2.7) and then recombined as a single band (see Figure 2.8). The elements of time are interchanged with each other, and the duration of elements is varied before being presented

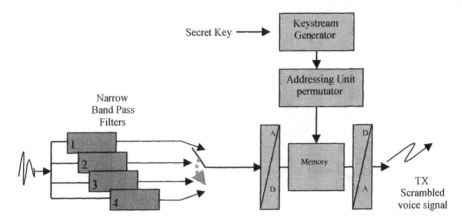

Figure 2.6   Frequency and time element scrambler (transmitter side)

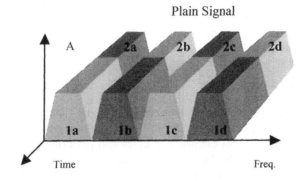

Figure 2.7   An analogue voice signal split into four sub-bands

for transmission. Both the band-splitting and the time element inversion are directly controlled by the secret key, the key generator and therefore the algorithm. Provided that the key is common to both the transmitting and receiving stations and there exists a method of synchronisation, the scrambled signal can be reconstituted by the receiving station or stations into the original voice message.

The band splitter separates the voice band into four parts, and the selector switch, according to the key data, selects the sequence of the band parts (see Figure 2.8), digitises the combined band and writes the corresponding blocks of digital voice data into a RAM memory device. The blocks of data are then read out of the RAM in a different order to the write in sequence, and hence the output will result in a combined frequency/time element scrambled analogue output. The digital output is once more converted to an analogue signal, which is forwarded on to the transmitter for sending.

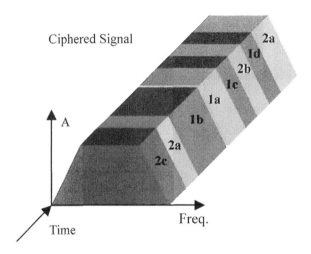

Figure 2.8   The scrambled band ready for transmission

Despite the inclusion of digital electronics in analogue ciphering or scrambling, whichever label is preferred, residual voice intelligibility remains to be heard on the transmitted signal. However, the latest methods have almost eliminated it and give a highly secure voice encryption that is difficult and costly to attack.

2.1.4 Digital Ciphering

Digital ciphering is quite different to the analogue process and can be achieved by stream or block ciphers. Digital stream ciphering seeks to break down the plain signal, whether it be digital or analogue, into single bit elements or bytes and performs a mixing by modulo 2 addition. The encryption can be bit by bit, by a key stream or by a block approach, but not for voice ciphering. In carrying out this procedure, digital ciphering seeks to emulate the famed 'one time pad' process, which achieves perfect ciphering by encrypting a message by using a random sequence of numbers that is never reused. The properties of a one-time pad are:

- A randomly generated key bit stream must be as long as the plain bit stream. This means very long keystreams for long messages
- The key bit stream can only be used once
- The key bit stream must be securely distributed
- The encryption itself is of a high quality, but the onus to preserve the security is on the process of key management.

Similarly, then, the requirements for digital stream ciphering are:

- The generated bit sequence must behave randomly, i.e. as a pseudo-random sequence
- It must be infeasible to derive the secret key (SK) from the keystream sequence.
- It must be infeasible to predict the key stream sequence without knowledge of the secret key
- The keystream must have a long period
- The transmitter and receiver must be able to synchronise on a bit-by-bit level.

2.1.5 Digital Stream Ciphering

Figure 2.10 illustrates the principle of encryption by digital stream ciphering, and the similarity with the one-time pad of Figure 2.9 is obvious. In the example above, the last four bits of the plain bit stream, 1010 (reading from left to right) are shown to be *modulo 2* added with four bits of the keystream 1001, to give the last four bits of the cipher stream, 0011. The decryption is carried out by a 'mirror image' operation at the receiver.

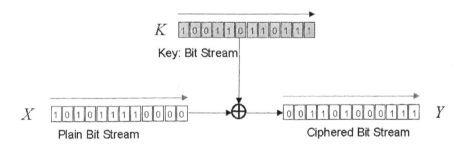

Figure 2.9   The principle of the digital one-time pad

The application to voice ciphering is modelled in Figure 2.11. The analogue voice signal is, or should be, taken directly from the microphone output, amplified and converted into a digital data stream. For the best security, the encryption should take place as close as possible to the source of the signal being protected and decryption, as close to the destination as possible. The digital stream is fed to a mixer unit in the form of an *exclusive OR gate* where modulo 2 addition of the data stream and the *keystream* is used to give an encrypted output known as the *cipher stream*. The key stream is the output of the algorithm built into the keystream generator, which is modelled here as a shift register, albeit an insecure one. The *cipher stream* is then presented to a modem and converted back into an analogue signal and prepared for the transmission whether by radio or telephone.

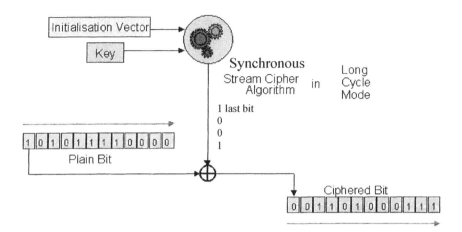

Figure 2.10   The principle of digital stream ciphering

At the receiving end, after filtering out noise, the received signal is amplified and converted from analogue into a *digital cipher stream* and fed to the mixer where modulo addition with a copy of the original *keystream* is carried out. The output of the deciphering module results in the extraction of the original data stream at the transmitter. A stage of digital-to-analogue conversion will reveal the original analogue voice signal, which would be amplified to drive a loudspeaker or earphone.

For successful reception in cipher mode, it is clear that the keystream at both ends of the transmission link must be the same. For this case to exist, both sides must carry the same secret key, i.e. symmetrical keys, the same algorithm and key generator as well as a secure and robust method of synchronisation. In Figure 2.9, it can be seen that if the 'rotation' of the key generators is not synchronised, the keystreams of the transmitter and receiver will not be the same, and therefore the message will not be deciphered correctly. A problem arises if the key generators were to start from the same position for every transmission or, in the case of a radio, on every operation of a push to talk (PTT). On such occasions, the keystream would be repetitive and therefore predictable. Using the same keystream sequence to encrypt messages would leak lots of information as the X-OR sum of two such ciphered messages would reveal the EX-OR sum of the corresponding plain messages. Also, if one message is known, the others would be revealed. To overcome this problem, an 'initialisation vector' (IV) that is generated by a random number generator is used to determine a new start position for the key generator. To ensure that both key generators operate in synchronism, the transmitting station sends the generated IV as an initial synch pattern, to the receiver whose key generator can then be brought into line with that of the transmitter. This is initial synchronisation, and it does not ensure synchronization throughout the whole message length. The subject of synchronisation is addressed further in Chapter 3.

The main properties of stream ciphers are that they give bit-by-bit encryption, they are fast, and they are unaffected by bit error propagation but bit slips, i.e. loss of synch, render deciphering impossible. So, a robust synch system is essential.

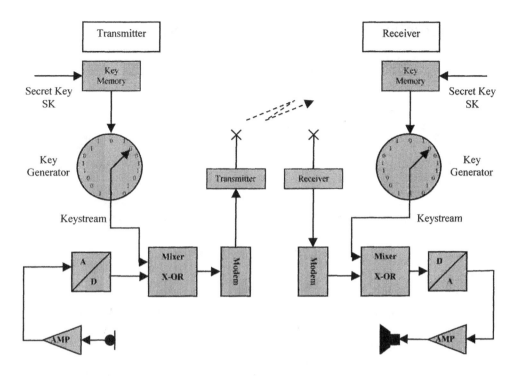

Figure 2.11   A model of the digital stream encryption of a voice signal

2.1.6 Block Ciphering

Block ciphers work on blocks of data rather than streams of data bits and are of typically 64 or 128 bit blocks. The general principle of block ciphering is shown in Figure 2.12, and for the sake of repetition, subsequently only the ciphering process of each mode is discussed, as the decryption process is merely the mirror image of encryption.

There are four main modes of block ciphering:

Electronic code book (ECB) gives a direct application of the block cipher and is only used for short messages, i.e. less than one block and also for disk encryption (see Chapter 10).
Cipher block chaining (CBC) gives chaining of the encrypted blocks and is used for longer messages, file encryption and secure IP.
Cipher feedback (CFB) is a block cipher that is used as a self-synchronising stream cipher. A self-synchronising cipher feeds back output ciphertext into the input, and therefore, the ciphering operation is a function of the previous bits. The decipher unit uses this information to synchronise its own key generator with that of the ciphering unit.
Long cycle mode (LCM) or counter mode is used as a synchronous stream cipher.

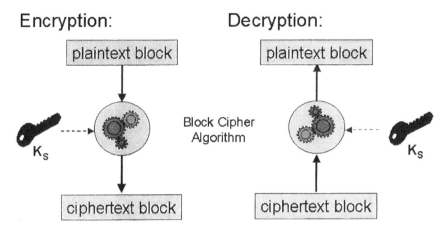

Figure 2.12   Basic block ciphering

2.1.6.1 Electronic Code Book (ECB) Ciphering

As can be seen in Figure 2.13, the *ECB* method encrypts a block of plaintext into a corresponding block of ciphertext, and the fact that the same plaintext is always encrypted to produce the same ciphertext implies that a code book of all plaintexts and corresponding ciphertexts can be assembled for each secret key, hence the algorithm name ECB. The method encrypts each block of plaintext independently, and this characteristic makes ECB a suitable algorithm for encrypting files that need to be accessed randomly with a view to making modifications to them, e.g. a database. As blocks of plaintext do not usually come in convenient packs of 64 or 128 bits, padding is required to fill the plaintext blocks up to the regulation size.

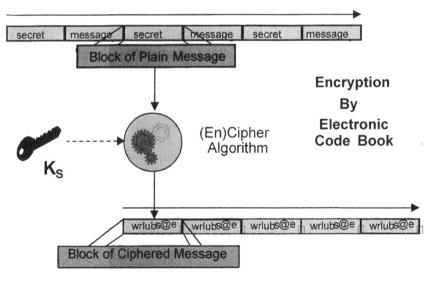

Figure 2.13   Electronic code book encryption

The main drawback of ECB is that an analyst can launch a statistical attack on a ciphertext based upon the fact that messages usually have an underlying common structure such as addresses, dates, sender's and receiver's name, etc., and to make matters worse, these are usually repetitive features. Given this information, the analyst is able to piece the jigsaw puzzle together and eventually develop the whole picture or codebook and thereafter to decipher any message encrypted by that specific key.

A second drawback is that if a bit error in the ciphertext is decrypted, then that whole block of deciphered plaintext will be corrupted. What is worse is that if a ciphertext bit is lost or gained (i.e. bit slip and bit insertion, respectively), the deciphering process is no longer properly aligned to the block boundaries. This problem also exists with both EBC and CBC modes but is only important when channels are likely to present bit slips or insertions.

A further weakness is that in a structured message such as a form, an attacker can modify a block of data to his benefit, without having access to the ECB key. Hence, further protection of authentication is needed.

On the positive side, however, the same key can encrypt more than one message, and the speed of operation is relatively fast.

2.1.6.2 Cipher Block Chaining (CBC) Ciphering

CBC encryption, as shown in Figure 2.14, introduces a feedback operation into the encryption process. In this manner, each encrypted block is fed back through a buffer register and is used in the XOR addition of the subsequent block. It follows that each encrypted block is dependent upon all of the other blocks that preceded it, hence the term 'chaining.' This mode is also vulnerable to analyst attack as identical messages will produce the same ciphertext, and repetitive message formats will present a 'way in' to the attacker. To overcome this weakness, an IV is generated by a random number generator and is used to fill the first block of

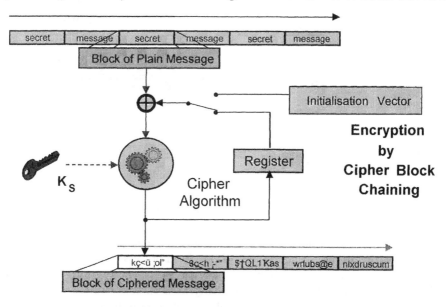

Figure 2.14 Cipher block chaining ciphering

ciphertext to randomise the initial state, thereby ensuring that each plain message is encrypted uniquely. The IV is of no security value other than its randomness properties, and so it is possible, indeed necessary, to transmit it to the remote station in plain, in order to synchronise the encryption/decryption process. As with *ECB*, padding is required to fill incomplete blocks.

The loss or inadvertent addition of a bit in the cipher text is destructive and results in a total loss of synchronisation of the encryption/decryption process as the message's block contents are shifted by the error bit. It is therefore essential that the frame structure of a *CBC* mechanism be preserved during transmission, as would also be the case with *ECB*.

Whilst plain message structures are usually hidden by *CBC*, the vulnerability to block modification by the intentional changing of a single bit and even block addition remains, and hence the integrity of the message has to be further protected. CBC has the positive features of providing fast encryption, the ability to cipher more than one message with the same key, and it exhibits the randomisation of the block input through the XOR gate.

2.1.6.3 Cipher Feedback Mode (CFB) Ciphering

Whereas CBC and ECB are restricted to ciphering 64 or 128 bit blocks of data, CFB (see Figure 2.15) can be used to encrypt smaller blocks such as bytes of data as found in ASCII characters. Following the random sequence of the IV input into the XOR gate, the ciphered byte is fed back through the register to perform a modulo 2 addition with the corresponding plaintext byte in the XOR gate. In this manner, the ciphertext output depends upon all the preceding plaintext.

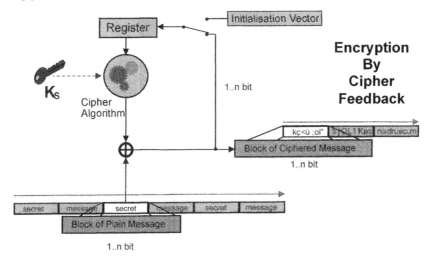

Figure 2.15   Encryption by cipher feedback mode

Once again, a bit error in the plaintext affects the entire ciphertext, but as with CBC, the error is reproduced as the original error bit by the decipher process. An error in the ciphertext will corrupt subsequent bits as it is deciphered and passed around the feedback loop until the error bit is lost from the feedback register. Similarly, CFB exhibits a robust synchronisation characteristic, and a lost or extra bit in the ciphertext will result in the loss of one byte of plaintext at the receiver before the cipher re-synchronises. A bit/byte error in the ciphertext will, after decryption, generate a burst of bit errors as long as this error bit remains in the feedback register. So, typically, the corresponding bit/byte is wrong. There is also a burst of the length of the feedback register, depending upon the feedback mode. In this case of bit slips or insertions, it usually re-synchronises after the correct transmission of 'n' bits, where 'n' is the length of the feedback register. Other positive factors include the concealment of plaintext message structures, randomised input into the cipher process, more than one message can be encrypted by the same key and the speed of operation is fast.

2.1.6.4 Long Cycle Mode (LCM) Ciphering

LCM has the same properties as the *counter mode*. As can be seen from Figure 2.16, this is a method of using a block cipher in a synchronous stream cipher mode. It is the only mode that exhibits no error propagation in case of error bits, but it is vulnerable against slips/insertions and may therefore need extra synch information.

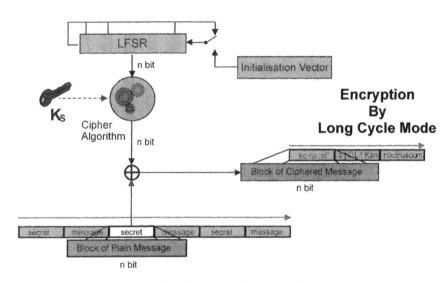

Figure 2.16   Encryption by long cycle mode

2.1.7 Summary

The main difference between stream ciphers and block ciphers is based upon their applications and the quality of the channel. In any case, as implied by the fact that only the encryption processes are described above, stream ciphers and block ciphers form the two categories of *symmetrical algorithms* of which the the Data Encryption Standard (DES) is the most famous, or infamous as some might say.

Stream ciphers are more readily implemented in hardware, whereas block ciphers are easier to implement in software packages and lend themselves to the encryption of computer data, which are essentially transported as blocks of information on data buses.

2.2 Algorithms

It is quite difficult to draw sensible comparisons between *symmetrical* (such as DES) and *asymmetrical cryptography* (such as the RSA® algorithm, Rivest, Shamir, Adleman), as each has its own distinct features and certainly very different applications. Symmetrical algorithms are far more suited to data, including voice data, encryption than are asymmetrical or 'public key' algorithms as they are far faster in operation and less susceptible to certain ciphertext attacks. However, when one considers the threats discussed in Chapter 1 and some of the applications in other chapters of this text, it is clear to see that symmetrical encryption is in itself a total solution for data protection. The technique certainly offers an efficient solution to the threat of message confidentiality, and it does infer authentication to some extent, but it does not ensure message integrity. In contrast, public key cryptography, which is undoubtedly slower and requires more computing power, offers a host of security features that symmetric algorithms cannot. Hence, as shown in Figure 2.18, public key cryptography is more suitable for key management than for the basic encryption of payload data. The intelligent combination of the two techniques into a cryptographic system produces a strong package of information security. The differences in application are highlighted in Table 2.1. MACs and hash algorithms are discussed in Section 2.3.3.

Table 2.1   Classes of algorithms

Algorithm	Authentication	Confidentiality	Integrity
Symmetrical	No	Yes	No[a]
Asymmetrical	Yes	Yes	No[a]
Message authentication codes (MACs)	Yes	No	Yes
Hash algorithms	No	No	Yes
Digital signature algorithms	Yes	No	Yes
Key agreement algorithms	Yes/no	Yes	No

[a] Depends upon the application.

2.2.1 Symmetrical Cryptography

The basic principle of a symmetrical key system is shown in Figure 2.17, where it can be seen that the essence of symmetrical encryption is that both the transmitting and receiving stations must have the same secret keys (SKs). They must also have the same algorithms, and subsequently, the keystreams in both the transmitter and receiver must also be the same. Symmetrical encryption relies upon the secure distribution of that common key to both locations. Indeed, the common key must be distributed to all parties of that network. It can be seen by the examples of Chapter 10 on 'Fax Networks', that complex network hierarchies using symmetrical keys can be organised to control communications by the suitable distribution of the common keys. It is the actual distribution of symmetrical keys throughout a network, by means of a 'secure channel' that creates the greatest headache for the security manager. The main danger is that anyone gaining access to the keys during distribution has

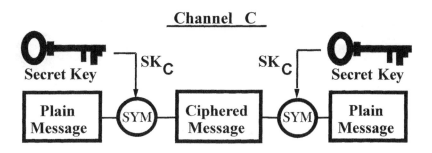

Figure 2.17   The principle of symmetric key systems

full access to the data that they are supposed to protect. Hence, key distribution must be by means of a 'secure channel' whether it be a physical or logical channel (see Figure 2.18). This subject is discussed in greater detail in Section 2.5.2.

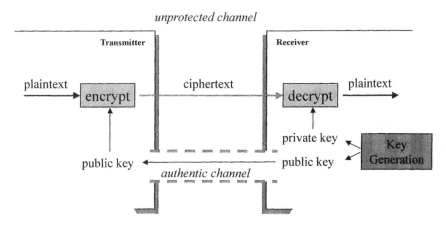

Figure 2.18   The theoretical channels of an asymmetric encryption system

2.2.2 Asymmetrical Cryptography

In contrast with symmetrical cryptography, asymmetrical algorithms operate using two keys or, more correctly, a key pair (see Figure 2.19). The keys are known as the *private key* and the *public key* and hence the term *public key cryptography*. Either key can be used for encryption or decryption, but unlike the symmetrical keys above, the key value or data of each is different from the other.

In the RSA® system, data encrypted by the *private key* (KX_u) of a key pair can only be decrypted by the *public key* (KY_u) of that pair and vice versa. The pair of keys is generated together and therefore related. It is essential to note that despite having this relationship, that even having access or knowledge of the public key, it is infeasible to compute the value of the private key. Therefore, there is no threat to the system by publishing the public key just as one might publish one's postal address, but the private key must be kept secret at all times. Herein

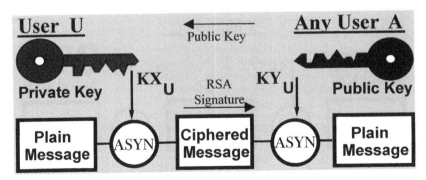

Figure 2.19   The basic principle of RSA® and public key encryption

lies the essence of RSA® public key cryptography whereby a ciphered message can be authenticated by the fact that if it is decipherable by the public key, the private key of that pair must have ciphered the message. Hence, the recipient of the message can be certain of the origin of that message. With the reservation that any holder of the public key could eavesdrop messages encrypted by the related private key, not only is the message confidential (within the group) because of the encryption, but it is also authenticated and, at the same time, non-refutable, i.e. the sender cannot deny that he or she was the author of that message. This contrasts distinctly with symmetrical operations where the encrypted message is confidential between the encrypting and the decrypting parties, but anybody else having the common key could have sent the message and yet deny sending it. Conversely, the recipient of the message cannot be certain of the message source. As will be seen in later chapters, there are ways to cope with this problem by the thoughtful use of key management, but there is no denying the fact that asymmetrical cryptography has certain advantages over symmetrical process in this respect. The downside is that asymmetrical algorithms require greater computing power and are considerably slower in the ciphering process than the symmetrical counterpart. There are also problems with error-prone systems such as in voice applications. Hence, public key encryption is rather unsuitable for messages of typical message length, but when combining the two modes, message confidentiality, authentication and integrity are theoretically assured. The most common use of asymmetric algorithms is in the:

- Signing of small '*hash*' values for digital signatures
- Encryption of secret keys used in symmetric algorithms
- Key agreement of secret keys between communicating partners

With little doubt, the most readily available public key package is the Pretty Good Privacy (PGP®) of Phil Zimmerman, and the present-day versions are easy to implement and play around with. The user is able to use a symmetrical key that is ideal for encrypting sensitive files stored on hard disk and to generate their own asymmetrical key pair for message security. The latter, following the principles mentioned above, requires the user to publish their public key whilst maintaining the secrecy of their private key. Over a period of time, one collects the public keys of all fellow users and a network of users grows in an informal manner rather than as a formal structured entity, controlled by a key manager. Possession of the public keys, on a 'key ring' allows the sender to cipher messages knowing that only the owner of the private key of that particular pair will be able to decipher the message, yet they will be able to decipher with their private key all of the messages sent and encrypted by his public key.

The researchers, Whitfield Diffie and Martin Hellman produced the first public key algorithm, but with key distribution in mind rather than straightforward message encryption, and it is in key management where asymmetric algorithms find their niche.

2.2.3 Hash Algorithms

As introduced in Chapter 1, hash algorithms provide an abbreviated check of a long, plain text message and are usually used as an *Integrity* check. Basically, a long string of data of variable length is input into the algorithm, which then produces a hash value of fixed length at the output. The one-way function makes it infeasible to work backwards from the hash value to reconstruct the original string, and it is also infeasible that any two input strings can generate the same hash value. Any changes to the input string will produce a different hash value at the output, thereby giving a check on the integrity of a message. Typical hash functions are MD 2, 4 and 5.

Hash functions, however, do not provide file or message confidentiality, although, when used with a secret key encryption, they can give authentication to a message.

2.2.4 MACs (Message Authentication Codes)

These are message encrypted integrity checks similar to hash functions, except that MACs are used with a secret key to provide both authentication and integrity checks, whereas hash functions, on their own, do not provide authentication. The MAC is most useful as a one-way hash function, although it may by used in a bi-direction mode. In the one-way mode, it involves the knowledge of a key by the author and the person wishing to verify the message or authenticate a file. As a consequence, the fact that a key is involved also gives some virus protection. A virus can infect a file and generate a new hash code, but as the virus cannot know the value of a key, it cannot generate a new MAC. Therefore, the author of a file will see that the file has been altered.

The disadvantages with MACs are that they are slower than digital signature algorithms, and as they are carried as appendixes to a file or message, they therefore add slightly to the overheads.

2.2.5 Digital Signature Algorithms

Digital signatures use public key algorithms to sign documents (see Figures 2.20 and 2.21). The message author carries out a hash function on the plain data of the message along with the date and time of that message to produce a message digest. When using the RSA® signature algorithm, the hash function produces a 'hash code' that is encrypted by the author's private key to produce a unique digital signature that is attached to the message in question.

The receiver verifies the authenticity and integrity of the message by running the verification algorithm with the received message, signature and the public key as inputs. The algorithm output is an indication that the signature is either valid or, in the case of the message being modified, invalid.

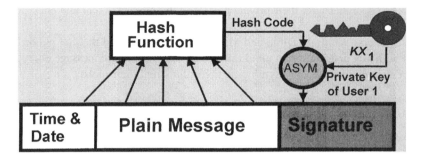

Figure 2.20   Digital signatures–generation

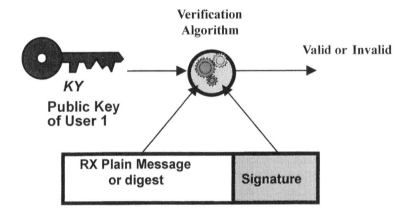

Figure 2.21   Digital signatures–verification

2.2.6 Key Agreement/Exchange Algorithms

The oldest public key algorithm is the 'Diffie–Hellman' algorithm. This algorithm is more useful as a key agreement tool rather than for direct message encryption. The main handicap with such algorithms is that they suffer from the 'man in the middle' problem. An eaves-dropper, monitoring the key agreement exchange, can capture the components exchanged in that key agreement process. This illustrates the fact that the authenticity of public keys is essential, and it is particularly obvious since new public keys are exchanged during each communication session.

More details and applications of MACs, hash functions, and digital signatures are discussed in Chapters 5, 11, and 12.

2.2.7 Summary of Comparisons Between Asymmetric and Symmetric Algorithms

- Asymmetric algorithms offer enhanced computational security by virtue of their different properties but cannot really be compared with symmetrical algorithms because of their differences.
- In large, fully connected networks, asymmetric encryption normally requires fewer keys

and less key data to secure a network than symmetrical encryption. In small, closed networks, symmetric systems can be just as practical.

- Asymmetric algorithms more readily allow separation of authenticity and confidentiality.
- Asymmetrical algorithms offer the facility of digital signature integrity.
- Asymmetric encryption is more often used for secret key distribution than for message encryption. Hybrid systems use a symmetric, randomly generated session key for the message encryption, but it is the asymmetric system that protects the session key.
- Encryption by asymmetrical keys is slower and requires more computer power than symmetrical encryption.

2.3 Goodbye DES, Hello AES

In October 2000, the search for a successor to DES came to an end. The replacement deemed to be the best of the candidates offered was the AES, the Advanced Encryption Standard submitted by Vincent Rijmen of the Katholieke Universiteit, Leuven in Belgium and Joan Daemen of Proton World International.

Originally produced by IBM and accepted as the standard by the now named NIST (National Institute of Standards Technology) in 1975, DES has been assaulted by every aspiring cryptanalyst and as early in its life as 1977, the writing was on the wall as to how long the algorithm would survive unblemished. As computing power advanced at a meteoric rate, the time required to carry out a brute-force attack to completion on DES became very manageable, and towards the end of the last decade, it fell to a period of about 5 days. Obviously, DES's time as a secure device had come to an end, and it is with much relief amongst the interested fraternities that a new tool will be permeating the industry. The application of triple DES as a method of extending its useful life improved the situation somewhat by applying the algorithm with two or three 56-bit keys, but Triple DES is slow.

Compared to DES with its woeful 56-bit key length, the AES, with a 128-bit key, offers a much higher security against a brute-force attack, and there are options to use longer keys up to 256 bits. The new algorithm is a block cipher and can be implemented in hardware and software in a wide number of environments such as smart cards, gate arrays and PC software. This ability was one of the characteristics that gave it superiority over its competitors, MARS, Serpent, RC6™, Twofish and the winner, namely Rijndael.

The long demise of and growing suspicions about DES stimulated the cryptographic industry to produce numerous alternative algorithms, and it will be interesting to see what effect the arrival of the AES has upon the rest of the industry. It will take some time, though, before it infiltrates the cryptographic niches. No doubt, competition for security markets will increase, and equally without doubt, every cryptographic seat of learning will turn its attention to uncover any as yet undetected flaw in the product. What transpires remains to be seen, but it is all part of the evolution of cryptology, and communications security will benefit from the interest and attention. The constant battle between cryptographer and cryptanalyst takes another turn. DES is dead, long live the AES?

2.4 Fundamentals in Key Management

The selection of a suitable algorithm for a secure network is without doubt a major issue, and most organisations concerned about their data security are particular about their final choice. However, even the most secure algorithm is practically worthless if it is not supported by efficient key management. In fact, key management is the Achilles heel of many a secure communications system. To attack an algorithm by cryptanalysis requires a considerable effort in terms of manpower and computing power, and usually involves a massive outlay of cash to fund a variety of attacks. In a highly secure system, this may be the only way in to gain access to an organisation's secret data, but history shows and statistics reinforce the fact that the most productive method of attacking a secure communications system is to influence the system's personnel and exploit weaknesses in its management. If one considers Table 2.2, it soon becomes clear that in order to break a key by 'brute force' or 'exhaustive key search', i.e. the testing of all key possibilities in order to come up with the one that will successfully decipher a message encrypted by a known algorithm, a considerable investment is required. Rather than spending millions on analytical tools to gain knowledge of, say, a 128-bit key, which is statistically impossible to attack within a useful period, it is much easier, and considerably less expensive, to exploit the failings of human nature. Conversely, it is naïve of an organisation to spend fortunes on the 'Rolls-Royce' of cryptographic systems and little on its management and management personnel. Much publicised events of recent history have underlined that a secure communications system will be attacked at its weakest point and is undoubtedly the human element whether it be purely operational deficiencies or the darker element of subterfuge. The purpose of key management is to reduce the threats of either of these vulnerabilities to a bare minimum and to process secret keys in such a manner that it is transparent to the user and the network. This section addresses these issues. Key management concerns the following aspects of the key lifetime (see Figure 2.22).

An integral part of key management is the training of management and operating personnel. This aspect is treated as a separate subject in Chapter 14.

Table 2.2   Estimates of time required to attack keys by exhaustive key search

Key length (bits)	Key variety	Tests/sec./ computer	Number of computers	Used time
40	1.1×10^{12}	10^9	10^3	1.1 s
56	7.2×10^{16}	10^9	10^3	20 h
80	1.2×10^{24}	10^9	10^3	38,000 years
128	3.4×10^{38}	10^9	10^3	1.1×10^{19} years
128	3.4×10^{38}	10^9	7×10^{9}[a]	1.5×10^{12} years

[a] World population.

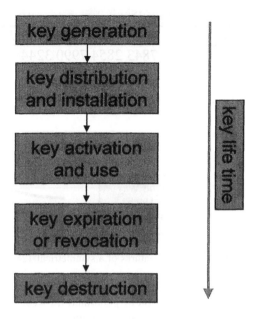

Figure 2.22   The key lifetime

2.4.1 Key Generation

Secret keys are usually generated either by the user or by a security manager, and whereas there are occasions when keys other than those generated by specialised management centres are necessary, the vast majority are ideally generated using a random number generator. The desire for randomness in keys is based on the fact that random numbers have no predictable sequence, and therefore, they are ideal for use as unpredictable secret keys.

The manual generation of keys is fraught with problems, the main problem being that through one's formative years, intentionally or otherwise, one is trained to think in patterns including numerical patterns, so much so that a statistical analysis of humanoid random number sequences readily underlines the trait, especially when more than a few keys are required to be generated. Manually producing hundreds of keys is not only tedious and stressful but also hands out an open invitation to the analyst to 'come and get it!'

There is a multitude of methods used for key generation, from seeds and pass phrases as found in PGP®, to those reliant upon the varying period between keystrokes on a computer keyboard and from the real random sequences derived from electronic/thermal noise sources such as P/N diodes and resistor elements. Whilst pseudo-randomly generated keys are acceptable for many applications, pure randomly generated keys are much sought after by the manufacturers of high-quality secure communications equipment. A quality management centre will include the option of allowing manually produced keys and the offer of random number keys. Any key centre worth its salt will have a built-in test procedure to check the quality of the generated key data and the ability to reject anything other than the genuine article. PGP® allows the user to generate key pairs from a pass phrase and the program checks the quality of the pass phrase before allowing the generation of the key pair. A choice

Key ID	Key Data	Signature	
01200	1873 9906 4057 2081 8425 1104 5095 0189 7744	1FA8	
01201	2107 4308 6837 0016 7843 2854 7990 3244 6518	595B	

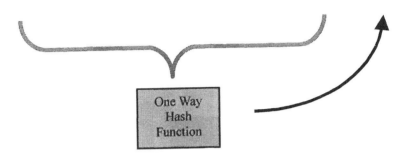

Figure 2.23   The main features of a symmetrical 128-bit key

of key length is offered, and the application of generated the keys, i.e. for RSA® or Diffie/ Hellman (El Gamal variant) along with a digital signature standard (DSS) is available. The generation of asymmetrical keys is more complex and takes somewhat longer than symmetrical keys.

Alongside the final key data in Figure 2.23, or value as they are often referred to, is a *footprint* or *signature* and a *Key ID* or *label*. More often than not, the key ID is a free choice and is used simply as an identifier for that key. It may well be convenient to link the ID with a period of validity. A footprint or signature has a different function, which is all important to the key manager. A *key signature* (see Figure 2.24) is computed by a '*one-way function*' that hashes the key data and produces a unique output byte of typically four or five hex characters.

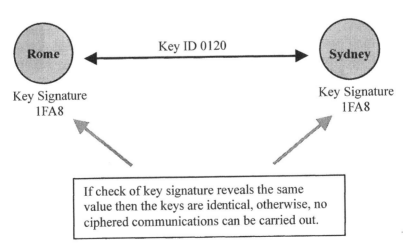

Figure 2.24   The application of key signatures

The *signature* is an extremely useful tool enabling the manager to check that the keys entered into one encryption device are the same as those entered into a corresponding, remote station. It is a key checking function whose divulgence over a public medium carries no significant security risk, as the signature is a product of a *'one way function'* the process of which is, by definition, infeasible to reverse. Bruce Schneier, in his excellent 'Applied Cryptography', likens the output of a 'one-way function' to the structure of a heavily smashed dinner plate in that it is easy to smash the plate but somewhat more difficult to put it back together again in its original condition. The signature becomes even more important if, amongst the options of a key generating management centre, there is the chance of generating the keys anonymously, in the sense that their values are not displayed. In this way, nobody has access to the key data, and hence one of the greatest threats to key security is removed. Instead, the verification of efficient key distribution relies upon the harmless knowledge of the signatures. In any case, it is far better for all keys to be generated by and distributed from a central control or management centre, where the quality and confidentiality of the keys can be controlled.

Key length is a subject that troubles many a prospective security system buyer, and from Table 2.2, one can see the reason why. The longer the key, the greater is the security against an 'exhaustive key search'. DES has a 56-bit key and is apparently vulnerable within about 20 h of attack, whereas a 128-bit key with a variety of 10^{38} can resist revealing all of its possibilities for rather longer, i.e. greater than the age of the solar system! So, it is no surprise that there is a consensus of opinion that DES has had its day and that 128-bit algorithms and above are the order of the day for high-security applications. So, why not aim for a higher security by using 256-bit or even greater length keys? The cryptologist would argue strongly that the marginal gain in key variety would be offset by the time cost in producing, testing and using these longer keys just to extend the brute-force time beyond what is already an incomprehensible period. For the uninitiated, the following example serves to illustrate many of the features of key generation discussed above:

The keys used are 32 digits, 4-bit/digit keys, which constitute a 128-bit symmetrical key. The key IDs are often free to choose and may well represent the validity period of the key. The time required to complete an exhaustive key search is calculated as follows:

The Key variety is given by

$$\text{Digit range}^{\text{no. of digits}}$$

i.e. 10^{32} where 32 decimal characters are used.

If the processing power available allows 1,000,000,000 key checks per second, then the time to check all possibilities is

$$\frac{10^{32}}{10^9} = 10^{23} \text{ s}$$

which approximates to 10^{15} years to cover all possibilities.

Permitting hexadecimal characters in addition to decimal characters extends the period, as does increasing the number of digits in the key value.

2.4.2 Key Storage

Other than during distribution, there are two aspects of key storage. One is in the management centre, and the other is in the actual ciphering units at their remote locations. The risks involved with a management centre are normally fewer than elsewhere as the centre should be located in a secure area and therefore in a controlled environment with access limited to only those who 'need to know'. Yet, the very fact that the management centre will probably carry information of a complete network makes it a prime target for attack. The situation is often exacerbated by the fact that secure areas are very prone to the weaknesses of routine controls that breed lethargy and vulnerability. Therefore, in addition to strict access controls to the location, sensitive data such as keys that are stored in the management centre should be stored in a secure manner. This means that they should be encrypted on a centre's hard disk by a database encryption key, and as even the best computers are known to fail from time to time, a complete network database should be stored on a back-up drive. For extra security, both the main drive and the back-up drive should be removable so that they can be stored in a secured safe when not in use.

At the distributed stations, access to key data should be forbidden to the user and impossible for the assailant to gain knowledge of. A hierarchy of passwords is useful but not sufficient on their own to protect the key data from attack. The answer to physical attack is the application of a tamperproof security module or, where the construction and application permit, a removable storage facility such as a PCMCIA PC card.

Another essential facility concerned with key storage is an emergency clear option to clear keys and other sensitive parameters quickly and easily, should the need arise. Having said that, the clearing of keys should not be so easy that it can be done accidentally.

A further requirement is that of a back-up battery to support key memories during times of temporary power failures or the movement of a cipher machine from one office to another. However, that is not to say that a device should be allowed to carry keys during shipping or transportation. This is far from the truth, and under such circumstances, an emergency clear operation should be carried out prior to dispatch in order to remove all sensitive parameters. An encryption unit should never be allowed to carry cryptographic data whenever it is in the hands of a third party.

2.4.3 Key Distribution

Of all the tasks facing the key manager, this is the most exacting for large networks, and it is almost certain that for the larger networks, especially those geographically displaced, a variety of delivery mediums and tools are necessary. Experience has shown that even in the best of circumstances, a certain percentage of distribution efforts will fail, at least in the first instance, and that alternative tactics need to be kept at hand to deal with such occasions. In the worst case, keys will be lost and perhaps key distribution devices along with them, and on such occasions, the response should be to assume the maximum threat, i.e. that the enemy has both the keys and a cipher machine in which to use them. Therefore, an immediate redistribution of new keys and key change must be carried out. The key manager must foresee such problems and be prepared to overcome all of them quickly if his network is not going to fall into disrepair and compromise.

There are three main families of distribution tools available to the administrator:

- Key transport devices (KTD)
- Down line loading (DLL) or over the air re-keying (OTR)
- Paper delivery with manual entry

Key transport devices (KTDs) include:

- Key kuns
- Laptops
- Smart cards
- Memory cards

    - Magnetic stripe cards
    - RAM chip cards

With the advent of a nefw generation of smart cards and improved card security techniques, perhaps the days of the key gun and such hardware devices are numbered. The typical key gun is heavily constructed with tamper-resistant circuitry protected by illegal entry switches to clear the key memory when any effort to gain access to the interior is made. Further protection is effected by passwords, which might include a hierarchy levels with different user rights for each. As with most password-controlled devices, key guns need to be protected against a 'brute force' attack on the access password by deleting keys after typically three erroneous log-in attempts. Various interfaces with the encryption device and the key generation centre are available, and a good idea is to design the key gun as a passive device as far as output control is concerned. Some such units use an infrared clock signal from the host devices to read out the key data rather than rely on internal control.

Laptop computers as distribution devices obviously have their applications, particularly when a large amount of security data needs to be downloaded. The programming of frequency hopping parameters into a squadron of aircraft might be one such case. Otherwise, laptops are expensive, relatively slow to operate compared with a dedicated machine, and are not ideal for transportation.

As discussed in Chapter 11, the very limited characteristics of magnetic stripe cards leave the smart card and RAM chip card as the most ideal forms of key distribution devices, and of these, the smart card is the favourite owing to its on-card processing facilities. The basic structure of a smart card is illustrated in Figure 2.25.

DLL is the process by which key data are sent over the transmission medium to remote encryption units. The idea of transmitting keys over a telephone line or radio link has been a subject of great debate within the secure communications industry. There are many proponents of the policy and still some sceptics on the security aspect of trusting 'public' media with this essential management task. Without doubt, DLL is the most convenient method of key distribution, especially over wide spread networks such as a Foreign Affairs embassy fax network. It has a distinct advantage of being almost totally controlled by a central body and with little reliance on other personnel. A sophisticated management centre will have automatic logging procedures for all DLL operations, and failures will be highlighted for immediate attention.

Naturally, nobody would advocate the transmission of plain key data, nor in fact, wherever possible, the distribution of plain key data in any form. The answer is to encrypt the Secret Keys with a key encrypting key (KEK) or, as it is often alternatively named, a key transfer/ transport key (KTK), as shown in Figure 2.26.

The encryption of key data by a KTK is not limited to DLL or OTAR operations but can equally be applied to key guns and distribution cards.

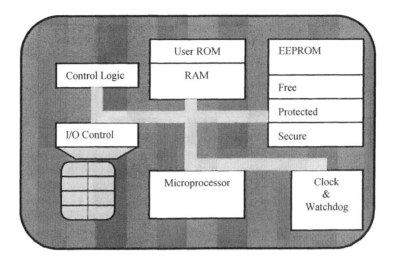

Figure 2.25   Smart card architecture

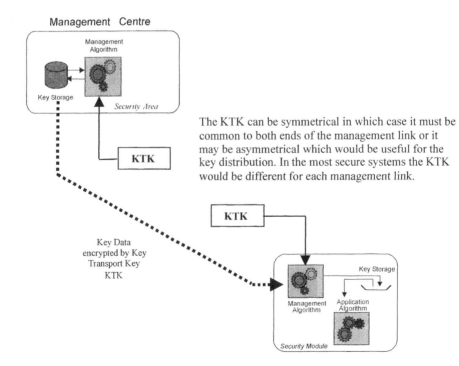

The KTK can be symmetrical in which case it must be common to both ends of the management link or it may be asymmetrical which would be useful for the key distribution. In the most secure systems the KTK would be different for each management link.

Figure 2.26   Down line loading with KTK protection

2.4.4 Key Changes

Of all questions related to key management, by far the most common and difficult to answer is 'How often must we change keys?' There is no simple answer to this question, and the decision of when to change keys is a balance of a number of factors, though the underlying theme being one of 'as often as possible' should always be remembered.

The factors involved include:

- The threat status
- Strength of the algorithms
- The ease with which keys can be distributed
- The amount of communications traffic
- The number of keys in use
- The known or suspected number of attempts to monitor communication links

The threat status means whether the 'organisation' is at peace with little imminent possibility of conflict, in which case, key changes can be less often, or when a threat of aggression exists and stimulates a more frequent change. When, on a war footing, a whole new ball game exists, and generally speaking, key changes should be made as often as feasible. Relatively speaking, a peaceful climate might require a monthly key change, the existence of a real threat, a weekly change, and in times of conflict, a daily change, but not forgetting the other factors listed above.

If an organisation is using an algorithm that is common to many users or one of less than 128 bits, then it is advisable that key changes be more frequent than with a strong proprietary algorithm.

Whilst it might seem that a daily key change would be the ideal solution to maintaining security, it makes little sense if the keys cannot be distributed throughout the network effectively. Too many changes in rapid succession can result in some stations being isolated from the main network, and in times of stress, this is the last thing that is needed. A three-key system is a useful method to overcome the vulnerability of a network during key changes. The system is demonstrated in Figures 2.27 and 2.28.

The system allows the Tokyo and Singapore Present key sets to be used for ciphering transmitted messages, but both Past and Next key sets can be used only for reception. When a key set change takes place, a new set, e.g. January, is loaded into the Next table, the December set is moved into the Present table, and the November set is moved into the Past table with the October set being lost.

The existing situation (for the purpose of this exercise) is that Singapore Station has missed a key change and hence is missing the January set and is still using the November set as its active transmitting set, whilst Tokyo uses December set for transmissions.

In the case where Tokyo transmits to Singapore, the key selected from its table has the label Singapore and Signature 4495. At the receiving end, Singapore will check through its Present set and will fail to find the correct signature. In a single-key system, no ciphered communication would be possible, but in the three-key system, Singapore will search through both its Past and Next sets to look for the key with signature 4495 and will find it in the December set, and therefore, it will be able to decipher the message from Tokyo. Conversely, when Singapore transmits to Tokyo, the former will use the key labelled Tokyo, signature 2FC4 in its Present set (November). On receiving the message from Singapore, Tokyo will search

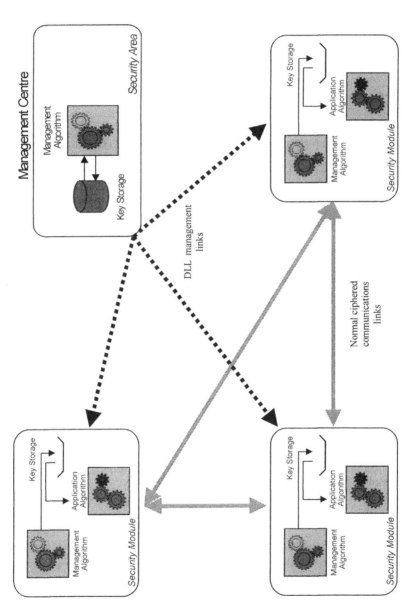

Figure 2.27 Key distribution from a management centre

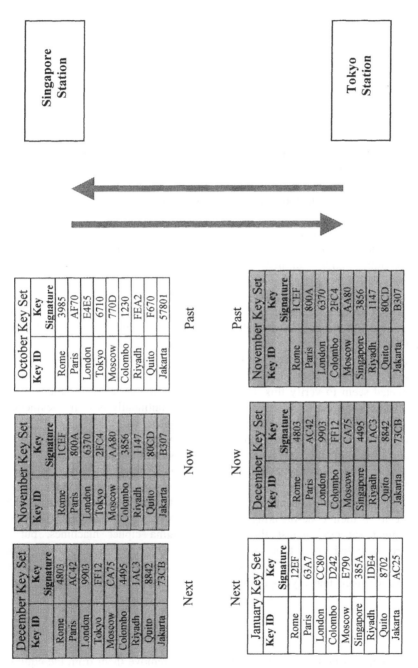

Singapore Station

October Key Set (Past)

Key ID	Key Signature
Rome	3985
Paris	AF70
London	E4E5
Tokyo	6710
Moscow	770D
Colombo	1230
Riyadh	FEA2
Quito	F670
Jakarta	57801

November Key Set (Now)

Key ID	Key Signature
Rome	1CEF
Paris	800A
London	6370
Tokyo	2FC4
Moscow	AA80
Colombo	3856
Riyadh	1147
Quito	80CD
Jakarta	B307

December Key Set (Next)

Key ID	Key Signature
Rome	4803
Paris	AC42
London	9903
Tokyo	FF12
Moscow	CA75
Colombo	4495
Riyadh	1AC3
Quito	8842
Jakarta	73CB

Tokyo Station

November Key Set (Past)

Key ID	Key Signature
Rome	1CEF
Paris	800A
London	6370
Colombo	2FC4
Moscow	AA80
Singapore	3856
Riyadh	1147
Quito	80CD
Jakarta	B307

December Key Set (Now)

Key ID	Key Signature
Rome	4803
Paris	AC42
London	9903
Colombo	FF12
Moscow	CA75
Singapore	4495
Riyadh	1AC3
Quito	8842
Jakarta	73CB

January Key Set (Next)

Key ID	Key Signature
Rome	12EF
Paris	63A7
London	CC80
Colombo	D242
Moscow	E790
Singapore	385A
Riyadh	1DE4
Quito	8702
Jakarta	AC25

Figure 2.28 The three-key system of key management

through its tables to eventually find the correct key in its Past set, and therefore, successful deciphering will be possible.

It can be seen, then, that the use of a three-key system can introduce some tolerance into a network where key changes are difficult to synchronize, and this will allow a station missing a single-key change to be able still to communicate securely until it is finally updated. However, should such a station miss two key changes, it will not be able to find a correct key to use.

A ciphered link carrying a lot of message traffic is more prone to cryptographic analysis than one carrying only light traffic, as the more often a key is used, the more vulnerable it becomes as any attacker is able to gather more information about the data on that link, making successful analysis more probable. Hence, it is essential that busy links receive more regular key changes than a lightly loaded link. Traffic analysis, which is not usually difficult to carry out, almost certainly will divulge the major stations or HQs of a network suggesting to the eaves dropper that these links are more worthwhile trying to compromise than less busy links, a fact that underlines the fact that more frequent key changes are necessary.

A network that uses one or just a few keys is also more vulnerable to attack, and there-fore, more frequent changes are necessary to maintain the network integrity. The same can be said for a communications link that involves a remote station located in enemy territory or, say, an embassy in an aggressive nation's country. During the Cold War between the Soviet Block and the Nato Alliance, the US embassy in Moscow would certainly have been a prime target for communications attack by the Russians, and the same can be said for the Russian embassy in Washington. No doubt, key changes would have been made frequently by both parties.

To summarise, a cryptologist concerned with a brute-force attack on their keys, bearing in mind the worked example in Section 2.3.1, would say that key changes are never needed, and the figures quoted seem to verify this. However, history and experience paint a different picture, and the consequence of failing to update keys continually is well documented in Khan's excellent read for security managers in 'Codebreakers'.

During the Second World War, the Japanese never broke a major American code, and in experiencing such difficulty, they believed their own ciphers to be largely invulner-able, experienced difficulties with key distribution and therefore made too few key changes. On numerous occasions, they were forced to run old ciphers in parallel with new ciphers to protect the same messages, a fact that greatly aided William Friedman and his US team in their code-breaking efforts. Similarly, in the European conflict, the British had great success against the German Enigma cipher machines, but their successes were frustrated every time the Germans changed the Enigma's rotor settings. These are lessons not to be ignored by today's communications security managers, and it should be acknowledged that cryptographic analysis is not the sole method of obtaining key data.

2.4.5 Key Destruction

It may seem a simple issue to destroy outdated keys, to remove them from the equipment and archives. However, the policy adopted depends very much on the type of commu-nication method and encryption. Where voice encryption is the case, there is nothing to lose by destroying dated keys totally with little need to archive them as the message has already become redundant. It is a different story when data messaging is involved as it is quite likely that computer data will be stored perhaps indefinitely, and therefore, the keys used for encrypting that data must be either retained for the useful life of the data or replaced by deciphering the data and then re-ciphering with new keys. Care must be exercised here, otherwise stored encrypted data would be lost forever if the related keys were destroyed.

2.4.6 Separation

One fear that is common to most organisations purchasing ciphering equipment is the suspicion that the producer may well have a trapdoor access to their client's security. This is a well-founded suspicion, especially when considering all the publications on key ESCROW policies being promoted by the US and British governments, the smoke generated by the never-ending speculation of the role of the NSA in the design of DES and the rigorous control on algorithm export licences. There is no doubt that many national security agencies would welcome the opportunity to gain a foothold on the secrets of both individual and national communications, and therefore, it is naïve to think that some manufacturers are not pressured to provide such niches in their crypto-graphy. In the end, when everything is taken into consideration, the decision to purchase one particular product or another is based on trust. Trust is something that is built up on reputation, experience, customer relations, the services offered by a company and also the sort of environment in which a company is working, its roots so to speak. Amongst the services that may be provided by the astute manufacturer are a number of innovations included to help allay the client's fears. These are generally known, or referred to, as 'Separation' in that they provide opportunities for the client to exercise some control over his algorithm and so distance themselves from both the manufacturer and any sponsoring authority.

The first level of separation comes from the basic generation and use of the secret keys. The second level is to be found in the use of a 'higher level key'. These are very useful keys in that a hierarchy can be created allowing the customer to both isolate himself from outside organisations and provide some internal separation. This facility is essential when, say, a nation's armed forces are operating not only in their own environment but also in an international environment that involves joint exercises or missions with organisations such as NATO or other alliances. Naturally, the application could equally be political. Figure 2.29 illustrates a possible scenario, and further application examples are demonstrated throughout the book.

A more secure separation is achieved by each customer having their own custom-designed algorithm implemented into his encryption devices. One step further, which allows the flexibility of that shown in Figure 2.29, is to adopt two algorithms, one for national use only and one for joint communications with international bodies. Table 2.3 illustrates a possible key hierarchy of a Home Affairs Ministry communications structure with 'EURO-FORCE' facilities.

Perhaps, the ultimate separation is achieved by allowing the client to either use their own algorithm or to modify the algorithm prepared for them by the manufacturer. This strategy provides many challenges to the manufacturer, not least in maintaining some control over

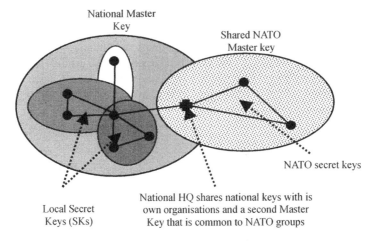

Figure 2.29   The use of master keys in constructing operating groups

the algorithm. This is a critical point as unregulated access to the algorithm can easily render it fatally damaged. Consider Figure 2.30, which models a key generator. The structure of the generator formed by a shift register, albeit extremely simple, is according to a connection diagram, i.e. the algorithm. The shaded blocks represent the basic unit with feedback links from connections A & B on the shift register, and the result of the operation of the clocked register is that a key stream is produced in such a fashion that it will not show any evidence of repetition for an extremely long period, typically years of continuous operation. This is known as the key stream period, and it is a major characteristic of a

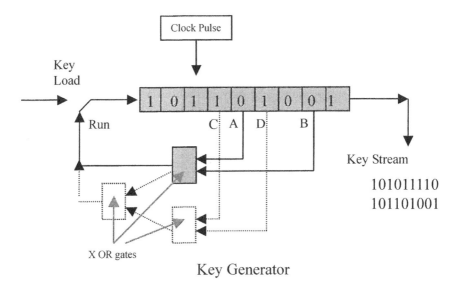

Figure 2.30   Separation by client modification of the manufacturer's algorithm

Table 2.3 The application of shared algorithms

	National Home Affairs Network	EUROFORCE Network
National Algorithm {	SK 001	
	SK 002	
	....	
	SK 100	
Shared Algorithm {	SK 101	SK 101
	SK 102	SK 102
	....	....
	SK 200	SK 102

key stream encryption device. In the example shown, the key stream period will be very short, i.e. in the order of a few tens of clock pulses, and insecure even if the shift register were very much longer, but it serves merely to illustrate the principle. Now, if the manufacturer permits, the client is able to modify the structure of the generator by adding the dotted connections C & D and components such that the key stream will be specific to that customer and their key inputs. However, the manufacturer must not know what exact modifications have been made, but they will need to take some measures to prevent the inadvertent demolition of the algorithm. It does not take much imagination to appreciate that if the wrong type of connections were made by the unwitting client, the key stream could be so seriously compromised that its key stream period might be drastically reduced. Therefore, it is in both interests that client and manufacturer work closely together to produce a satisfactory solution.

Separation is a very sensitive subject and one that is not discussed readily in the public domain, but it is an important factor to be considered, hence its modelled inclusion in this text.

2.5 Evaluating Encryption Equipment

Strictly speaking, this section could be easily written as a complete chapter, or even as a book on its own, but that is beyond the specification of this text. However, having made that point, most of this chapter is pertinent to the assessment of encryption equipment, and this final section, a summary of it.

Perhaps the most obvious factor to consider when assessing encryption equipment is that of the manufacturer's algorithm. Whilst it is certainly a fundamental factor, and its selection is a major strategy decision on the part of a procurer, this is certainly not the sole point to be discussed. In fact, in assessing any equipment or system, it is the whole package that must be considered.

2.5.1 The Main Points of Evaluation

Equipment evaluation criteria:

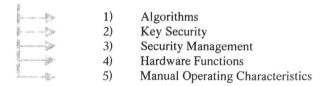

1)	Algorithms
2)	Key Security
3)	Security Management
4)	Hardware Functions
5)	Manual Operating Characteristics

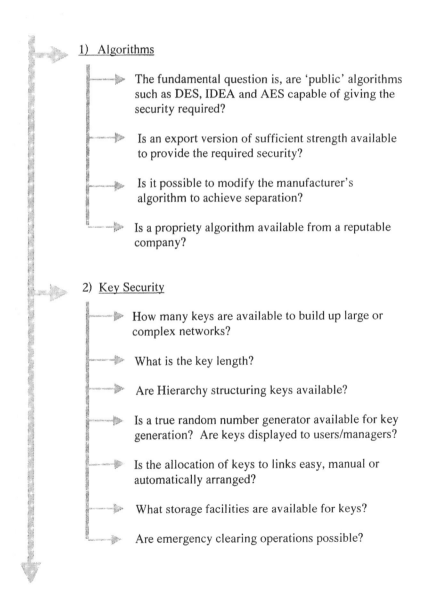

1) Algorithms

The fundamental question is, are 'public' algorithms such as DES, IDEA and AES capable of giving the security required?

Is an export version of sufficient strength available to provide the required security?

Is it possible to modify the manufacturer's algorithm to achieve separation?

Is a propriety algorithm available from a reputable company?

2) Key Security

How many keys are available to build up large or complex networks?

What is the key length?

Are Hierarchy structuring keys available?

Is a true random number generator available for key generation? Are keys displayed to users/managers?

Is the allocation of keys to links easy, manual or automatically arranged?

What storage facilities are available for keys?

Are emergency clearing operations possible?

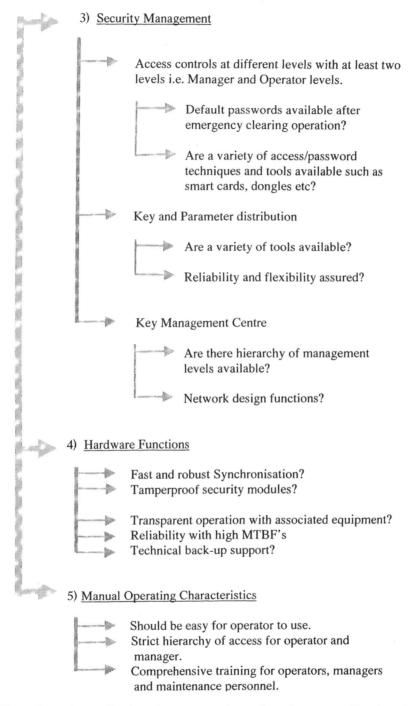

3) Security Management

Access controls at different levels with at least two levels i.e. Manager and Operator levels.

Default passwords available after emergency clearing operation?

Are a variety of access/password techniques and tools available such as smart cards, dongles etc?

Key and Parameter distribution

Are a variety of tools available?

Reliability and flexibility assured?

Key Management Centre

Are there hierarchy of management levels available?

Network design functions?

4) Hardware Functions

Fast and robust Synchronisation?
Tamperproof security modules?

Transparent operation with associated equipment?
Reliability with high MTBF's
Technical back-up support?

5) Manual Operating Characteristics

Should be easy for operator to use.
Strict hierarchy of access for operator and manager.
Comprehensive training for operators, managers and maintenance personnel.

Throughout the application chapters, specimen data sheets are offered to the reader as assessment material in support of the theme of this section.

3

Voice Security in Military Applications

The main threats to voice communications in a military environment are eavesdropping, impersonation and illegal access to the medium. The degree of threat depends very much upon the political climate of the region involved and whether or not any parties are taking an aggressive stance towards a neighbour. In times of peace, one would expect little, if any, interference in the day-to-day communications of a nation's armed forces. That is not to say that monitoring of the radio medium by a friendly neighbour does not take place, for it certainly does. Even in the best and most stable of alliances, suspicions and fears exist, arousing a 'need to know' of the inner thoughts of one's partners, in order to facilitate political positioning and the development of subsequent military strategies.

There is much evidence of communications naivety in many parts of the world, and as a result, many a nation's sensitive data have been freely available to those who care to listen. The relaxed attitude that often prevails on the subject of communications security in peacetime enables eavesdroppers to build up a big picture from fragments of information. This is largely gleaned from poor telecommunication security practice. Maintaining high standards of operation also helps an organisation to gain an insight into the communications practice and identify weaknesses of the target community. This knowledge serves as a useful foundation on which to build communications analysis techniques that will serve the eavesdropper well at some later date. Should a situation of unease arise, with the possibility of developing into a moderate threat or something more severe, such practice gives a head start in resolving any conflict. The near catastrophic stance that Britain took before the onset of conflict with Germany in 1939 was somewhat tempered by the fact that the nation had the fledgling communications security infrastructure in place. Communications monitoring and analysis techniques were already quite well developed, and of even greater importance was the fact that this was a well-kept secret. The US, whilst not starting the Second World War in quite such a difficult state of preparation, were also very naïve in many respects, but as did their counterparts across the Atlantic, they too had a foundation on which a communications security strategy was able to flourish. The outcome is history and for all to see. It is therefore essential for an organisation to maintain its communications confidentiality even during peacetime and, what is more, to have its own independent, well-developed security. It is a comfortable position to be enjoying compatibility with allies when engaged on joint exercises, as we see with NATO and similar alliances and yet benefit from security separation for own-force communications. To maintain confidentiality at any time, encryption of communications is essential, and as threats grow, authentication becomes more and more a necessity. Once a relationship starts to deteriorate and the threat increases, the need to be sure about the origins and targets of messages becomes critical. Opposing forces increasingly take steps to degrade each other's communications efficiency and integrity as a position deteriorates. The final offensive, in communications terms, is not only to eavesdrop and confuse but ultimately to deny the opposition the very medium by which they communicate.

In a military scenario, 'stand-alone' ciphering devices might, on first consideration, seem to be an unnecessary complication in a security strategy. What they do introduce, however, is the option to implement proprietary encryption, rather than relying upon 'off the shelf' standards. Stand-alone encryption devices give an organisation the freedom to purchase their high-class radio transceivers from any available source without being influenced by the restrictive export restrictions of cryptology packages. The client is then at liberty to search the market for the high-security equipment of their choice, without being inhibited by government restrictions or undesirable sponsorship. For many armed forces, a customised solution for communications security is a very desirable strategy, and a 'mix and match' policy can result in an impressive system.

This chapter looks at three alternative strategies of preserving a military force's radio communications. The first option describes the application of a stand-alone analogue encryption device, as may be found on a naval high-frequency (3–30 MHz) (HF) radio platform. Whilst the security of analogue encryption was brought into question in Chapter 2, such devices do still exist, and it is interesting to compare the analogue and digital processes. Therefore, an application is included here. It is only recently that the problems of bandwidth and the dispersion of HF signals have been overcome by the use of modern vocoders and equalisation techniques. These advances have opened the door for digital encryption of HF voice transmissions, and today, we can expect to see encryption for this medium more in the style of that shown in Figure 3.6, i.e. digital, rather than the analogue form described in Figure 3.3.

The second part of the chapter looks at a digital alternative, also a stand-alone device, whilst the third part explains how cipher modules can be integrated into an existing 'off the shelf' radio transceiver, and finally, whilst the subject of frequency hopping is dealt with in some greater technical detail in Chapter 7, a somewhat different application is offered here. Battlefield radio operations have largely been for voice communications in the past, but the demands of modern military operations are demanding even more complicated and sophisticated packages that deliver both voice and data communications, hence a brief excursion into such techniques.

3.1 Analogue Encryption of Naval Long Range, HF Radio Communications

The essential difference between naval communications and those of other forces is that navies having units dispersed throughout the oceans of the world frequently need to rely upon communications over great distances (see Figure 3.1). The radio medium most suitable for this application is HF transmission or, in the case of submarine communications, even lower down the frequency spectrum. Of course, a ship's communications installation will

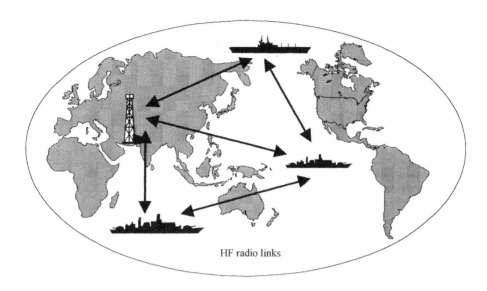

Figure 3.1   HF naval communications

include many radio options, i.e. the ultra-high frequency (300–3000 MHz) (UHF) and very high frequency (30–300 MHz) (VHF) bands for line-of-sight transmissions as well as long-distance HF communications. The HF radio installation of a medium-sized naval ship will look something like that illustrated in Figure 3.2. As one might expect, there are a multitude of possible configurations depending upon the type of ship, its size and the function that the vessel performs.

3.1.1 Ship Communications Operation

A ship's communications set up is defined by:

- The internal communications package.
- The external links to other ships or to shore
- A method of connecting both systems together

When integrating the different components, it should be borne in mind that there must be a separation between the non-secure and the secure communications operation. Whereas internal communications are usually through 'Tannoy' or telephony and often in plain, external ship-to-ship and ship-to-shore transmissions are most certainly in cipher mode. This is a typical example of *Red/Black* separation of radio systems, and great lengths are taken to ensure that the partition is not compromised by the inadvertent transmissions of plain messages or peripheral information that might be of interest to eavesdroppers. The interconnection of the various communication systems is shown in Figure 3.2.

The signals routing is as follows. Suppose the officer on the bridge wishes to call the galley. He can do this quite simply by using the ship's internal communications centre control, without passing through the matrix or patch centre, as it is otherwise known. However, when that officer wishes to make a plain voice transmission to another ship, perhaps of a

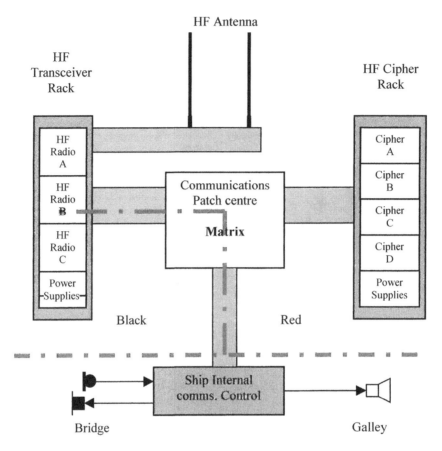

Figure 3.2 Signal routing in ship communications

commercial nature, from the same location, his call will be directed to the matrix, which patches the signal to a selected HF radio and subsequently to an HF antenna for transmission. When the similar situation occurs, but with a requirement to transmit in cipher mode to either a naval ship or base HQ, the voice signal will be patched first to an available cipher unit and then to an HF radio for transmission through a suitable antenna. The reverse situation applies when receiving a radio call, in which case, the destination of the signal can be either to the bridge handset earphone or to any other communications station on board the ship via the patch centre.

The system above would be paralleled for VHF/UHF installations, which makes for a complex installation, but one that is totally flexible and able to cope with adversity, largely by duplication. Another element not shown in Figure 3.2 is the possibility of making calls to or from the communications centre as well as the possibility of monitoring of calls at that point. In a sense, the centre can be thought of as a relay centre or as a terminal. This raises important considerations about security in that, if a confidential voice conversation is ongoing between the commander of the ship and that of another ship, steps have to be taken to prevent the illicit monitoring of such communications, by the staff located in the communications centre. The problem is that whilst, ideally, one would wish the cipher unit to be located as close as possible to the source of the signal, i.e. the microphone and conversely to the destination, i.e. adjacent to the speaker or earphone, it is neither practical nor financially feasible to install a cipher unit at each communications station on the ship. Therefore, discipline and practice come very much into play.

3.1.2 The Cipher/Scrambler Features

The typical stand-alone analogue cipher module, and its interfacing, is illustrated by the block diagram Figure 3.3. This drawing emphasises the functionality rather than the practical implementation that we would find on inspecting a state-of-the-art scrambler device. The modern application would certainly use a digital signal processor (DSP) for the band-splitting process.

In transmit mode, the plain voice signal is brought in from the microphone through a connecting board and an interface unit. The purpose of the interface unit is to process the plain input signal so that its characteristics are compatible with the internal processes within the ciphering operation. Similarly, the ciphered output signal must be prepared for passing to the radio transmitter such that the ciphered signal characteristics match those of the transceiver input. The situation in receive mode is very similar except that the input signal will be coming from the output of the radio receiver, and the characteristics of that signal must be processed so as to be compatible with the decipher process. The output of the decipher process must be conditioned to drive either a loudspeaker or an earphone. In addition to the voice signals, various controls, e.g. key selection, key clearing commands and data inputs such as key loads must also pass in through the connecting board and interface unit and be dealt with accordingly. Display data and alarms pass in the opposite direction, from the internal structure to the outside environment.

If the signal processing is somewhat obvious, then probably less obvious is the degree of input and output filtering and shielding that is required to enable the device to meet EMC and TEMPEST standards. This is especially important in communications centres when transceivers can be influenced by stray Black signals from various peripheral sources. This subject is discussed in greater depth in Chapter 1.

The main signal flow within the analogue cipher unit is as follows. The plain input signal from the microphone is a small analogue signal of a few millivolts amplitude, and therefore, the first process is to amplify this in the audio frequency transmit (AFT) amplifier by a factor of about +30 dB or more. This is so that it matches with the requirements of the unit's internal logic and digital coding. A provision is usually made for the option of plain operation, in which case, the voice signal will be switched through the TX (Transmit) amplifier, directly back to the connecting board and to the transmitter beyond. It may be that switching to plain operation will bypass the cipher unit completely, i.e. externally, but there may be a case for 'plain override' or 'fallback', in which case, the switching action must be within the cipher unit. In cipher mode, however, the voice signal is passed through a filter to limit the bandwidth of the signal to typically 3 kHz and then to two, or more, band-splitting filters. The purpose of this signal filtering is split the bandwidth of the original voice signal into two or more sub-bands, depending upon the actual device specification. The band-splitting is also described in some detail in Chapter 2. Once the individual band components have been established, they are converted into blocks of, say, 8 digits by an analogue-to-digital conver-

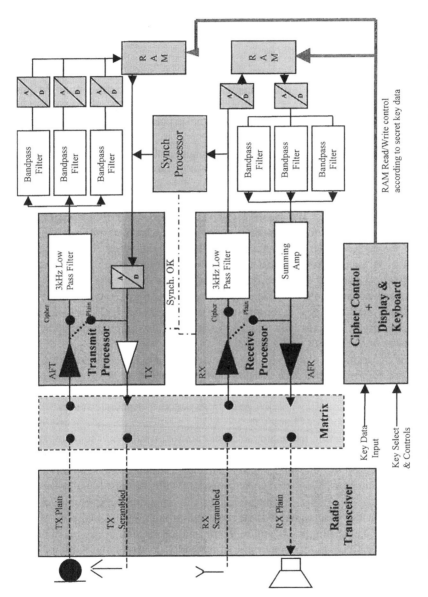

Figure 3.3 Block diagram showing the implementation of an analogue scrambler in a ship installation

ter. Working in digital form at this point facilitates more efficient scrambling of the bands. It is a much simpler task to do the scrambling this way, by interchanging digital blocks, rather than implementing the scrambling by purely analogue means. Each manufacturer has their own options as far as the scrambling process is concerned, and whilst some vendors simply transpose the sub-bands, others transpose, invert and compress the voice signal to give both greater security. This also removes excessive voice content (phonemes) from the transmitted signal. The digitisation also provides the opportunity to add the essential synch information that must be transmitted along with the voice signal so that it can be detected and extracted by the receiving cipher unit.

There are a variety of synch operating methods, but they usually incorporate an initial synch, a continuous pattern, or periodic synch bursts that will enable *late entry* into the ciphered communications as well as giving a robust synchronisation that resists interference and fading.

'Digital scrambling' can be achieved by writing the digital voice blocks sequentially into a RAM and then reading them out in a different order according to the cipher key selected to protect that communication (see Figure 3.4). Once the scrambling has been carried out and the synch information added to the transmit package, the digital signal must be converted back to an analogue form by passing it through a digital-to-analogue (D/A) converter. The scrambled, transmit signal level is adjusted in the TX amplifier to fall within the input range of the transmitter. As the transmitter is normally expecting the voice signal directly from a microphone, of a few millivolts amplitude, the output signal level from the cipher unit must be of the same order. Hence, the last amplifier stage TX, will be an attenuator having negative rather than positive gain, as might have been first expected. The scrambled voice signal is then fed to the transmitter for transmission.

In receive mode, received signal is passed from the radio receiver to the cipher unit input and amplified by the RX (receive) amplifier to a suitable amplitude according to the input characteristics of the cipher unit. Once again, a provision to switch from cipher mode to plain mode should be provided, and in this case, the received plain signal would be fed directly to audio frequency receive (AFR) amplifier and out to either the radio loudspeaker or earphone.

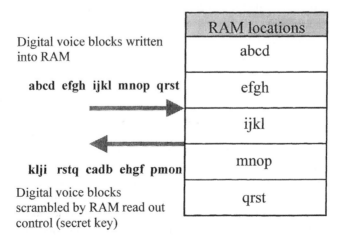

Figure 3.4 The principle of digital scrambling

However, in cipher mode, the complete voice signal is filtered to remove the synch information that was added at the transmitter. The remaining voice component is converted to a digital stream by the analogue-to-digital (A/D) converter and then passed through to the descrambling process. At this point, the digital stream is reassembled into blocks of data that are written into the RAM memory of Figure 3.4 and then read out in a sequence controlled by the same cipher key as was used by the sender for that communication. Provided that the key has the same value as that at the transmitter, i.e. symmetrical encryption or scrambling, and the cipher units are synchronised, the digitised blocks of voice data can be recovered from the RAM exactly their original sequence. The analogue voice signal is reconstructed by converting the digital blocks back into analogue form by the D/A converters. The analogue results are filtered to define the original sub-bands and then recombined to reconstitute the complete, deciphered voice signal at a summing amplifier. There is a final amplifier stage AFR, which is the driver stage for the local handset earphone or loudspeaker. Alternatively, the descrambled voice signal is passed back to the matrix from where it can be directed to any station on board the ship.

3.1.3 Synchronisation

As inferred elsewhere in this text, synchronisation of ciphering or scrambling operation is a very sensitive area of study, and individual manufacturer's techniques are almost as closely guarded as the encryption process itself. However, the function and some technical aspects of synchronisation are modelled in this section and in Section 3.2.

In Figure 3.3, it can be seen that the synch control is common to both transmit and receive sections. In transmit mode, a synch pattern is injected into the voice signal at the sender, during the digital scrambling process. The synch pattern is transmitted before the first voice data and subsequently as bursts between the ongoing voice transmission bursts. So, there is an initial synch followed by a continuous process constructed from the periodic bursts (see Figure 3.5). At the receiving end, the cipher unit continually monitors the input signal from the radio transceiver by filtering out the expected synch signal from the composite received signal. At this stage, the synch component is in an analogue form and is converted into a digital data stream so that it can be analysed for the actual synch content. Once the pattern is recognised as being the true, expected synch signal, the synch controller is certain that the key generators at each end of the transmission link are in step and can signal the go-ahead for de-scrambling to take place. The command signal to start the process is labelled as 'synch OK' in Figure 3.3, and it is only on this command that actual reception of a scrambled or ciphered takes place.

As the transmitter continues to inject synch information into the signal path for the duration of the transmission, sufficient information can be collected over a period of time for any other receiver having the same key to synchronise and join the link after missing the initial synch pattern. This is known as 'late entry' and is a common feature of modern radio security practice. The continuous 'bursting' of synch data also has a second function, which copes with the noisy radio environment. Multipath propagation and other sources of interference can result in the transmitted signal becoming corrupted and therefore resulting in a loss of synch. A reliable radio system must have a robust synch performance to maintain synchronisation. Therefore, the synch bursting during a continuous transmission can be used to 'top-up' the synch process so that the cipher unit can both tolerate a degree of errors and re-synch

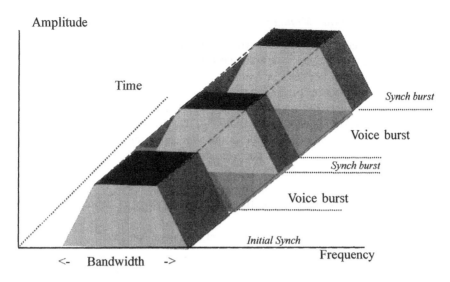

Figure 3.5   Synch bursting

quickly, should the link be temporarily lost. One method of exercising this control is described in Section 3.2.

3.1.4 Security Parameters

A typical device of this nature will have a secret key (SK) capacity of 10 keys, each probably consisting of 32 digital or hexadecimal characters, possibly defined according to the limitations of the keyboard or key programming device. This arrangement gives an ample key length of 128 bits for the cipher control of the scrambling process. Naturally, a strong preferably proprietary algorithm is desirable to give a long key stream period, along with the possibility of customized separation between the manufacturer and the client.

Access is by a password hierarchy, allowing operators to make simple parameter settings such as a change of key selection and yet permitting security officers to make more fundamental changes, e.g. key loading, signature checking and time settings for time authentication, etc. The latter should be included to overcome the threat of message reply or spoofing, as described in Chapter 1.

An emergency clear function should be implemented to allow for the 'zeroising' of all secret keys in case of the immediate danger of the ship being boarded. (History has shown that captured surface ships and submarines have often been the source of several cryptographic 'pinches', i.e. seizure of security equipment.)

Plain override is a function that, when enabled, will allow a receiving station to be automatically switched to plain mode, when receiving a plain message from a remote transmitter. The ideal system, as far as security is concerned, is to communicate in cipher mode only, thus forming a closed system. However, the ideal situation rarely exists, and there are occasions arising when plain communication is the only alternative. It is a useful option, but, if enabled, there must be adequate warning to operators in the form of visual and audible alarms clearly indicating that such a condition is exists.

3.1.5 Key Distribution and Management

Key distribution in naval applications involves careful planning for the future, as it is often the case that ships are away from their home ports for considerable lengths of time. Therefore, the update and loading of keys can be a problem. Long-term planning is essential with keys being generated and stored for the whole period of absence. They would be either programmed into courier devices such as 'key or fill guns', or carried by smart cards if they and their readers are available. Alternatively, they may simply be written on paper as part of the ship's sealed orders and stored in the ship's safe for opening and loading as and when required. The latter condition is far from ideal in the secure sense, and strict measures must be taken to preserve their confidentiality.

Key management can take many forms, depending upon the type of operation being undertaken, but given the limit of the 10 secret key capacity as defined above, Table 3.1 might be close to the desired option.

Table 3.1   Possible SK key management for a naval operation

Secret key ID	Key signature	Function	Validity
001	AE39	Common to all ships	January–May
002	5590	Red Sea fleet	January
003	56FB	Atlantic fleet	January
004	21C0	Pacific fleet	January
005	61D3	Ships–HQ	January
006	EF09	Common to all ships	June–December
007	5BED	Red Sea fleet	February
008	387A	Atlantic fleet	February
009	CC84	Pacific fleet	February
010	2F6E	Ships–HQ	February

3.2 Stand-alone Digital Cipher Units in Land-based Operations

A major requirement of army radio communications is 'line of sight' operation and to secure these communications against eaves dropping, and to ensure their authenticity, digital encryption is the norm. Figure 3.6 illustrates the basic principle of digital encryption of an analogue voice signal. This section is dedicated to the description of such a 'stand-alone' encryption device.

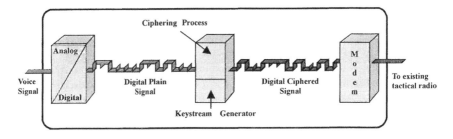

Figure 3.6   The principle of digital encryption

3.2.1 The Ground Force Scenario

In a battlefield scenario where situations are extremely fluid, commanders are in constant demand of information and the freedom to give out orders as and when needed. Naturally, training and preparation go a long way to making life as easy as possible for communications personnel, but to complement the human condition, equipment should exhibit the following major operational requirements:

- Reliability
- Flexibility
- Mobility
- Robustness
- Simplicity of operation

Unfortunately, in the history of recent military conflict, it is somewhat surprising, given the technology available to armed forces, to find that despite the manpower skills and bravery, technical failures have been frequent and costly. Secure communications in high-pressure environments, should exhibit a very high degree of transparency as far as the operator is concerned. Yet, that fundamental demand is quite difficult to achieve as telecommunications become ever more complex. Military radio equipment purchased today is all about integrated systems, as discerning buyers explore the market for the single solution device that satisfies the requirements of:

- Internet messaging
- Video and picture transmission
- GPS, situation awareness
- Point-to-point voice communications
- Key management and distribution
- Remote control capability
- Frequency hopping

Some of the points mentioned above have been dealt with in some detail in other chapters of this book, some merely inferred, but others go beyond the purpose of this manuscript and also beyond the author's experience. Therefore, this section, which could be the topic of a whole series of books, is confined to deal with the security of radio communications by digital encryption, its management and practice. It discusses devices operating in the VHF/UHF band of operations, namely a stand-alone backpack/vehicle mounted cipher module, a radio/integrated cipher module intended for backpack use and a brief excursion into the properties of a fully integrated multipurpose transceiver.

3.2.2 The Cipher Unit Features

The digital cipher unit shown in Figure 3.7 has many similarities with the scrambler unit in Figure 3.3, but there are also some fundamental differences. The major design constraints are laid down by the bandwidth availability of the radio channel. Naturally, the channel bandwidth in HF is less than that available for VHF/UHF channels, and so, whereas the scrambler described above is designed to operate within a 3-kHz band, a digital cipher unit will require a minimum of about 10 kHz of bandwidth. The latter figure enables the digital throughput of a

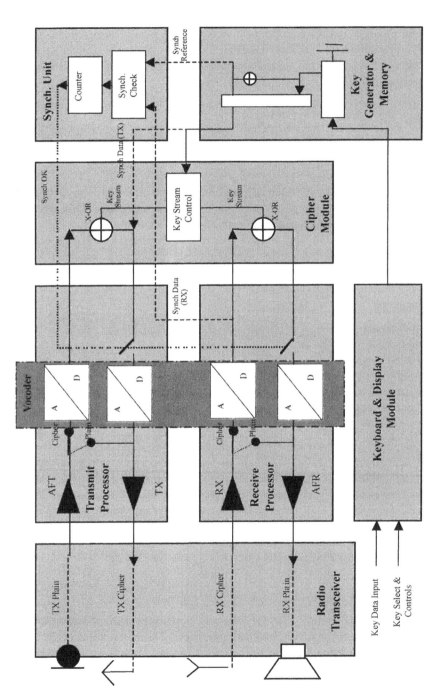

Figure 3.7 Block diagram of a digital voice encryption device

voice-ciphering device to operate typically between 9.6 kbits/s and 16 kbits/s with extra bandwidth space required for packaging overheads. These data rates can be accommodated by vocoders, and their use gives a tremendous boost to the quality of voice communications.

Otherwise, the main difference between the devices is that the digital ciphering operation is a pure stream ciphering operation using symmetrical encryption. The signal flow is as follows. In transmit mode, the analogue voice signal, originating from the microphone and being of a low amplitude, requires amplification by +30 to 40 dB, depending upon the type of microphone and any processes that are carried out by the radio. This gain is achieved by the amplifier AFT. There follows the cipher bypass option, which can accommodate the *plain transmission* should it be implemented. If this is the case, the plain audio signal is switched directly to the TX amplifier before being passed to the transmitter. A similar situation is to be found in the receive mode.

In cipher mode, however, the analogue signal is converted into a digital stream. A vocoder may be used in which various processes, including the A/D conversion, may be carried out. It is beyond the scope of this text to give a detailed description of vocoder operations, and the cipher process is only concerned with the digital output signal that is fed out of the vocoder processor, i.e. the voice components to be transmitted.

This digital stream is passed to the cipher module, where *modulo 2 addition* with the keystream is carried out in an Exclusive–OR gate (X-OR). The keystream is produced by the unit's key generator and is subsequently the product of the secret key (SK).

At this point, the initial synch preamble, originating from the synch unit, is injected into the signal path and precedes the digital cipher stream, which is converted back into an analogue signal in preparation for transmission. Once again, the output of the cipher unit must be matched to the input characteristics of the radio transceiver, and so the TX amplifier gain must be adjusted accordingly. It is usually the case that whatever positive gain is imported on the input signal by the AFT amplifier, it is compensated for by the the TX amplifier attenuating the signal by the same degree. Hence, a large degree of transparency of the cipher unit is presented to the external system, i.e. the transceiver. The TX ciphered voice signal is passed to the transmitter for transmission through a suitable antenna.

In receive mode, the RX ciphered voice signal is delivered to the cipher unit's input from the radio receiver and is amplified by the RX amplifier, such that the signal level is suitable for the A/D conversion. The result is the ciphered digital data stream, which is then fed to the cipher module for decryption. The decipher process once again uses the modulo 2 addition of the ciphered data and the keystream to extract the original plain voice data, albeit in digital form.

Providing that the synch OK signal has enabled the signal decryption, the deciphered signal is passed to the Codec for conversion into an analogue signal. The AFR amplifier boosts the recovered voice signal to drive either a handset earphone or a loudspeaker, which may be an integral part of the radio system. Further amplification would probably be required should an external speaker be included in the installation.

The complete process is dependent upon the successful acquisition of synchronisation between the participating key generators and assumes that the correct synch pattern has been detected before the voice signal was received. This is necessary so that the key generators will operate in step to produce identical keystreams. In order to achieve synchronisation, the contents of the synch preamble must be received error free by the decipher unit, and therefore, an error protection technique such as forward error correction (FEC) must be used in the transmission to ensure this. Failure to synchronise the cipher units must result in the inhibition of the received signal reaching the output, i.e. the audio transducer.

3.2.3 Synchronisation

One method of synchronising the key generators of cipher units is illustrated in Figure 3.8. When the transmitter is keyed into the transmit mode by a push-to-talk (PTT) operation, the synch sequence is generated by the synch register in the cipher unit. These data are then transmitted to the receiving station prior to any voice signal being sent. The receiving cipher unit then checks the input data it receives for a synch sequence.

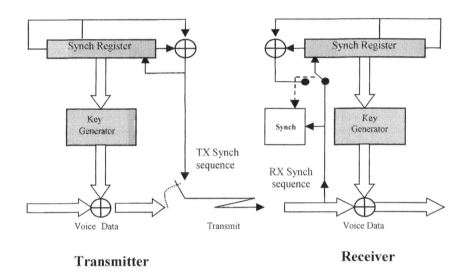

Figure 3.8   Synch sequence generation, transmission and detection

There are many variations on synch detecting techniques, and a second technique is illustrated in Figure 3.9. It works on the principle of searching through received data from a remote unit and checking it against a known pattern, i.e. the synch pattern, in the local unit. Once the pattern in question has been detected, synchronisation is achieved, and ciphering and deciphering can go ahead.

In the synch module of the local cipher unit, a known binary pattern, or a repeating pattern transmitted from the remote cipher unit, is loaded into the local shift register.

Two possibilities of pattern checking exist. In the first, there is a known synch pattern stored locally in each unit, and, on receiving a signal, the known pattern is loaded into the synch register and then compared bit by bit, with the stream of data coming from the remote transmitter. When the two bit streams are found to be identical, the received sequence can be acknowledged as being the correct synch sequence. This technique needs to be supported by the use of an 'initialisation vector' so that some degree of randomisation of the starting position of the key generators is achieved. During transmission, the IV must be protected against bit errors by an error protection mode such as FEC, otherwise synch cannot be achieved.

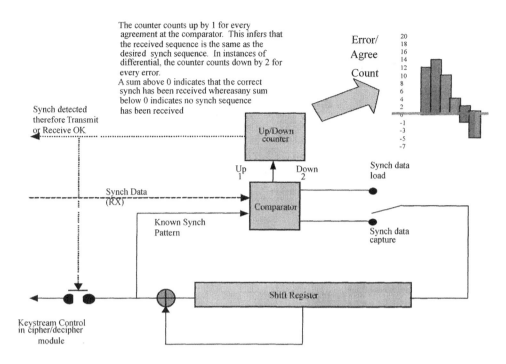

The counter counts up by 1 for every agreement at the comparator. This infers that the received sequence is the same as the desired synch sequence. In instances of differential, the counter counts down by 2 for every error.
A sum above 0 indicates that the correct synch has been received whereas any sum below 0 indicates no synch sequence has been received

Figure 3.9   One version of a synchronisation technique

Alternatively, a repeating known pattern is transmitted from the remote unit and subsequently received at the local end of the link. The local synch detector, not knowing where the sequence is in the continuous stream of incoming data, must continually check it until it detects the repeating pattern. In either case, at that point in time, both the transmitter and receiver have the same sequence available to synchronise their key generators.

The operation (see Figure 3.9) uses the comparator output to count the agreements between the two comparator inputs. It counts positively for every bit agreement, up to an upper limit, and counts down by a factor of two for every disparity. When the sum falls below '0', too many errors have been detected in the sequence being checked, and it can then be assumed that the input data are no longer the synch sequence. In this way, the synch criteria permit a number of errors, e.g. bit slips before it decides that synch has been lost, and so the method infers some robustness in a noisy radio environment. Without the system's tolerance, synch would be lost every time an error check occurred. Once synch has been agreed and providing both units have the same key installed and selected, then the key generators will generate identical keystreams at the transmitter and receiver. As the radio is a half duplex operation with a PTT control, there is no handshake exchange, and therefore no key agreement is possible. However, the output stage of the receiving cipher unit should be disabled by its failure to receive a 'Synch OK' control signal from the synch detect circuit or software.

One other common synchronising system is the recursion aided rapid acquisition by sequential estimation (RARASE), which is based on the use of a pseudo-noise in the transmitter and receiver with identical feedback logic, as shown in Figure 3.10. The method takes in the received sequence and loads it into the shift register whilst also providing one input to

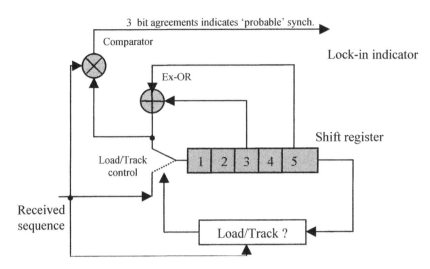

Figure 3.10   Principle of the RARASE synch. System in a receiver

the comparator. The shift register feeds back outputs 3 and 5 to an EX-OR function to produce a sequence when the register loop is closed. The EX-OR output bits are compared with the received input bit stream in the comparator, and when three bit agreements are made, the comparator signals that synchronisation is probable, and a trial tracking follows. If the tracking fails, the process is repeated until it succeeds, at which time, a synch ok signal indicates that both key generators in the transmitter and receiver are in step and that deciphering can commence.

3.2.4 Security Parameters

The following parameters are highly desirable in a military application:

- Default keys, with the possibility of enable/disable according to user profile.
- Robust synch system to contend with interference and difficult environment.
- Plain override with alarms, enabled according to user profile but very useful in case of loss of keys, but only to be used with extreme caution.
- Authentication of TX & RX units. Users must be certain about the identification of the remote party, and this can be achieved during a key agreement process along with pre-programmed cipher unit IDs that are exchanged during call set-up.
- Session keys, to add extra security by automatically generating a different ciphering key for every call, i.e. PTT operation.
- NMKs (Net Master Keys) can be used to define domains and/or achieve separation between clients and manufacturer.
- 10 SKs. The number of ciphering keys determines the flexibility in network topology design, i.e. key allocation to different groups and also facilitating the frequency of key changes.
- Time authentication, to eliminate the possibility of being threatened by replay.

- Emergency clear is an essential security feature that is used to erase all sensitive data. Automatic key selection according to the channel selected makes for ease of use in the field and restricts the possibility of erroneous call set-ups.
- Split algorithms give added security when used for client separation and also for performing different security tasks, e.g. encryption, key distribution and key storage.
- Manual input of keys. This is useful when keys have been lost or when normal key distribution by other techniques has failed.
- Over the air (OTA) or DLL key distribution. A very efficient method of key and parameter distribution from a management centre to a field station.
- Fill, or key guns, an efficient and flexible method of key and security parameter distribution, especially when used as a distribution package with other methods.
- Password access, an essential feature permitting pre-determined access level for users and managers.

3.2.5 Key Management

3.2.5.1 Monday Net

Figure 3.11 shows a typical basic network structure that may be used for a VHF/UHF operation. The keys are defined in Table 3.2. A network such as this is fine for short-term operations with a daily key change. The present loading for the cipher units is for Monday with a provision for a key change at a convenient time, to Tuesday's keys. Ideally, with a key change operation, individual units failing to make a change for some reason should have a

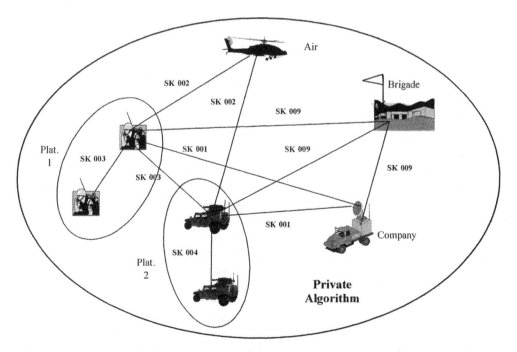

Figure 3.11   Network topology for MONDAY NET by private algorithm and secret keys (SKs)

Table 3.2   An alternative key use and management strategy for MONDAY NET

Secret key ID	Key signature	Function	Validity
001	1539	Common to Company HQ	Monday
002	4480	Ground/air support	Monday
003	5F35	Platoon 1	Monday
004	2890	Platoon 2	Monday
005	6433	Common to Company HQ	Tuesday
006	A12E	Ground/air support	Tuesday
007	690B	Platoon 1	Tuesday
008	386E	Platoon 2	Tuesday
009	CC57	Brigade HQ	Monday
010	62FE	Brigade HQ	Tuesday

mechanism for selecting a default key or falling back to the old key, i.e. Monday's key. As suggested elsewhere, the reason why a unit has failed to change keys should be investigated and user authentication confirmed. The three-key system largely overcomes missed changes. Of course, there are many variations possible on the same theme, depending essentially on the nature of the mission.

Being a single force network, i.e. a private net, it is not difficult to manage, but the situation becomes more complex when engaging in joint operations with the forces of another nation, as happened in the 'Gulf War' and 'Bosnia', etc.

The first problem is that individual nations will probably be in the possession of different cipher machines that are incompatible with others, and then either a rapid purchase of common ground equipment must take place, or communications must be held in plain mode: hardly an ideal situation, and one to avoided at all costs.

With joint force missions, individual nations will wish to communicate on a Global Net that is common to all participants of that force yet, at the same time, be able to maintain their own private net communications. In such cases, the NATO Net illustrated in Figures 3.12 and 3.13 offers the simplest solution. All parties of the NATO Net must have compatible encryption devices with the same algorithm installed. The separation is carried out by generating different secret keys, i.e. SK 1 for the private net and SK 2 for the Global Net. Naturally, the network structure can be expanded to give as many keys as are available for each of the subnets enabling each part to develop a reasonable secure network structure. The network is defined by the common algorithm, the number of SKs available and how they are distributed. The existing key distribution is shown in Table 3.3.

Some nations might feel uncomfortable about having to purchase a ciphering device that appears to have a universal algorithm and to share it with other forces that someday may be fighting from the opposite corner. Their reservations are well founded, and if that fear is at all acute, an alternative strategy that guarantees both private security, yet allows shared secure communications when the occasion arises, must be developed.

If sharing communications security assets with other forces does present a problem, there are two strategies to follow, which should offer some assurances to the sharing parties. The first, and perhaps least expensive, option is to purchase an algorithm that can be customized. When contemplating joint operations, all sharing parties must agree on an algorithm that is

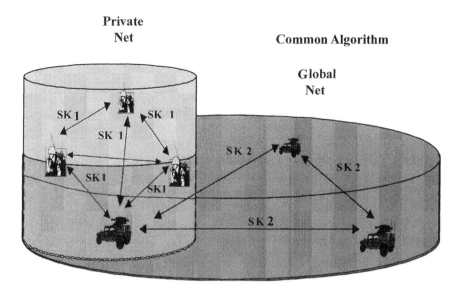

Figure 3.12 Joint operations with network definition by secret key management

common to all parties of the joint force. Any existing key must be first erased from the cipher machines before loading in the new joint key. Once this has been achieved, the rest of the common network structure can be organized by the use of secret keys (SKs). Once the

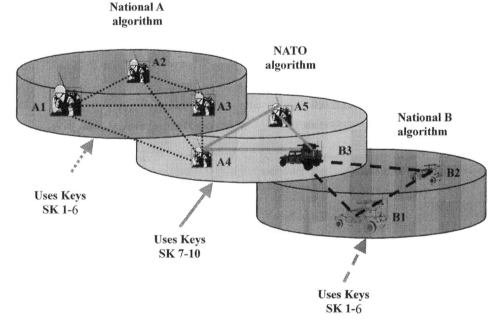

Figure 3.13 'NATO net' using a shared algorithm

Table 3.3   The key distribution for NATO Net

Secret key ID	Key signature	Function	Validity
001	1539	Common to GLOBAL	Week 1
001	1539	Ground/air support GLOBAL	Week 1
001	1539	Platoon 1 GLOBAL	Week 1
001	1539	Platoon 2 GLOBAL	Week 1
001	1539	Common to GLOBAL HQ	Week 1
002	A12E	Common PRIVATE	Week 1
002	A12E	Ground/air support PRIVATE	Week 1
002	A12E	Platoon PRIVATE	Week 1
002	A12E	PRIVATE HQ	Week 1
002	A12E	PRIVATE GHQ	Week 1

mission has been completed, the individual services can erase all of the key parameters that were used for the mission and re-enter parameters that are appertaining to their own, individual requirements, i.e. returning to their original set-ups. Whilst this enables full encryption of communications during joint missions, it does have the drawback that it precludes the simultaneous ciphering of communications with one's own group and the joint force. However, there should be some flexibility of the use of the SKs, and dedicating one or more of these to provide individual privacy is possible.

A more comprehensive, and almost certainly more expensive, solution is to search for a security provider who can deliver encryption equipment with two algorithms. This would allow one algorithm for private use and one for joint use. In this way, there is complete separation between all parties contributing to the joint force and yet full security integration is allowed for joint operations. This alternative is illustrated in Figure 3.13, and the key structure is shown in Tables 3.4 and 3.5.

Table 3.4   The key table according to National A's requirements

Secret key ID	Algorithm	Key signature	Function	Validity
001	National A	CD41	A1–A2	Private
002	National A	9865	A2–A3	Private
003	National A	FD67	A1–A3	Private
004	National A	00AB	A2–A4	Private
005	National A	B093	A1–A4	Private
006	National A	710A	A3–A4	Private
007	NATO	690B	A4–B3	Mission duration
008	NATO	386E	A5–B3	Mission duration
009	NATO	CC57	A4–A5	Mission duration
010	NATO	62FE	Spare	Mission duration

Table 3.5   The key table according to National B's requirements

Secret key ID	Algorithm	Key signature	Function	Validity
001	National B	5609	B1–B2	Private
002	National B	AE32	B2–B3	Private
003	National B	1095	B1–B3	Private
004	National B	FFCD	Spare	Private
005	National B	88FE	Spare	Private
006	National B	4067	Spare	Private
007	NATO	690B	A4–B3	Mission duration
008	NATO	386E	A5–B3	Mission duration
009	NATO	CC57	A4–A5	Mission duration
010	NATO	62FE	Spare	Mission duration

Consider Table 3.4. It is drawn up for the situation as far as a National A is concerned and has 10 SKs available for use. The first six keys are allocated to National A's private group, and it can be seen that in selecting those keys, the group's own algorithm is used for the communications over those channels to which the keys have been allocated. The function of the SKs is listed in the table, and it is through the designation of the keys that the 'A' network structure is defined. Naturally, the links defined are simply expressed as a model of a real situation, which would involve sub-groups according to the requirements of National A's security policy. It is important to note in particular the key signatures that have been generated from the SK data and the 'A' algorithms, and that they are different to those generated as private keys for the National B table. This confirms that the keys of the two private groups are incompatible. Even if the actual key data for each key were exactly the same in each case, i.e. the data for National A, SK 001 being identical to that of National B's SK 001, the fact that each uses a different algorithm ensures that the keys are incompatible. National B can also be seen to have six keys available for its own private use. These are detailed in Table 3.5, and the validity depends entirely on the individual group requirements.

In each table, and as illustrated in Figure 3.13, the NATO keys are common to both 'A' and 'B' groups, and the fact that the signatures are the same in each table indicates that both the key data are the same and that the keys (SK 007–SK 010) are using the common 'NATO' algorithm, i.e. they are compatible with each party. Therefore, complete separation between the individual groups is achieved whilst allowing a common facility, all in complete confidence that the secure communications of each separate party is totally confidential.

Once the topology of a joint mission has been defined, the next problem to deal with is that of key distribution and in particular the control of key changes. For pre-planned missions of short duration, most aspects can be dealt with reasonably, but for lengthy missions, headaches abound.

The ideal situation would mean a central controlling body formed from mission member representatives so that careful consideration of each individual party's privacy interests can be made. Coping with the politics of building a security management hierarchy will most probably be a major issue, but we must assume here that a suitable team is available to face the task. Whilst the distribution of each nation's individual keys will be a sensitive subject and of no simple matter itself to carry it out, the management of the NATO keys is likely to be a much more complex exercise. It is in circumstances such as this that the use of a key management centre (KMC) comes into its own and especially when it can provide for a multimedia courier distribution system. Another important facility, especially in these circumstances would be the option to generate keys from a random number source but to never display the actual key values to the user. In Chapter 2, I termed this operation 'anonymous key' generation. Using this technique, it would be possible for each national representative to generate the keys for their own organisation in a private manner and combine them with those generated as common NATO keys to produce a package of keys for distribution to their own forces. The keys can be written to courier devices according to the type of cipher equipment that nation has installed. In the case of incompatible hardware, it might be necessary to fall back to pare distribution with manual key entry into individual units. This would be the final option and would only be used where no alternative medium is available. The system is simply modelled in Figure 3.14.

In any case, these circumstances present an exacting challenge to security managers, especially when handling key changes over extended periods. It is one that requires strict central control and is a good subject to be included in joint military exercises so that efficient routines can be established and fine tuned as the application requires.

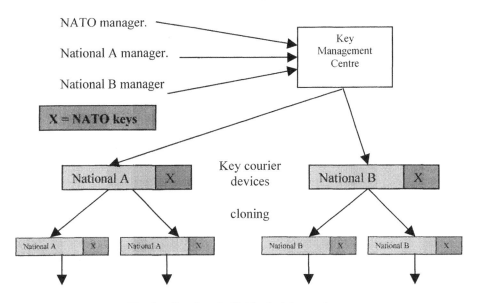

Key loading into individual cipher units

Figure 3.14   Key distribution system for 'NATO Net'

3.3 Radio Integrated Cipher Module

The idea of integrating a high-class cipher module into an existing radio transceiver is a very attractive consideration. Apart from the prospect of being able to shop around for a proprietary cipher unit, unrestricted by export restrictions that might well be the case when buying off-the-shelf equipment, there are many other advantages of implementing an integrated unit, including:

- Less weight for backpack operations
- No connecting cables
- Physical protection from being located into radio transceiver
- Transparent to the user with all control operations being linked directly to the radio's controls, e.g. SKs allocated to channels used and therefore automatically selected as the operator selects the respective channel
- Possible to compensate for lack of display data by incorporating this into audio alarms or commands through the audio output of the radio
- Low power demand

3.3.1 Typical Features

- Use of voice coders to give excellent voice quality and enhanced voice recognition. A vocoder is a device that takes a speech signal as an input and generates certain parameters from it such that when part of the speech signal is transmitted, it can be reconstructed at the receiver.
- Good key distribution tools, e.g. fill gun or smart cards + reader.
- Fallback to analogue plain operation in case of fault condition developing in the cipher module, thereby maintaining communications operability.

3.3.2 Cryptographic Parameters

These would include:

- Proprietary algorithm with strong algorithm and multi-key sets
- Key stream period of 200 years
- Key variety of 10^{38}
- Emergency clear 'zeroization' from radio control panel
- Late entry facility
- Fallback to a default key
- Plain override reception
- Automatic key change
- Time authentication to prevent message replay

3.3.3 Other Security Parameters and Features

The encryption process, synchronisation and key management will be similar to those highlighted in Section 3.2 and hence are not repeated here. Figure 3.15 illustrates the main components of just such a device.

The block diagram illustrated in Figure 3.15 gives an indication of the architecture of a compact cipher module. The major differences between this and other circuits, explained in this chapter, is that it is very much smaller and usually gives a far better voice quality than the

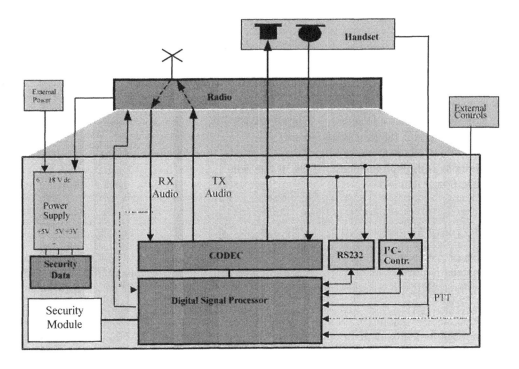

Figure 3.15   Cipher module integrated into a radio transceiver

previous traditional methods. It also encapsulates the ciphering process in a single application-specific integrated circuit (ASIC) chip that will most probably be a manufacturer-specific device. This of course makes unauthorized access to the security parameters and hardware algorithms much more difficult. A precision CODEC and some considerable computing power give rise to a sophisticated and versatile unit.

In transmit mode, the microphone signal is filtered to reduce noise and limit the bandwidth and is then converted into a digital signal, all by the CODEC. The CODEC output is fed to the DSP where the speech processing is carried out. DSP functions include:

• Signal filtering
• Convolution, i.e. the mixing of two signals
• Correlation, the comparison of two signals
• Amplification, rectification and/or transformation of a signal

It replaces all of the similar functions that were previously carried out by cumbersome analogue circuits. As far as secure communications are concerned, the main purpose of the DSP is to reduce the bandwidth of a transmitted voice signal by removing much of the residual voice information from the microphone signal and presenting the basic message contents for encryption and transmission. At the receiver, the residual voice patterns are added to the received, deciphered message, to give excellent voice quality that was not previously available. The ASIC design and structural secrets are very closely guarded by manufacturers, who are most reluctant to give much details of their component. This is not surprising given that the development of a custom-built ASIC involves considerable manpower and associated cost, not to mention the security aspects of the structure. However, its basic function is to take the filtered voice signal, cipher it, convert it from a parallel data stream and modulate it before passing the output to the transmitter board. Reception is the reverse process with the ciphered, receiver output being deciphered in the ASIC before the audio residual information is added to the received voice signal. This is processed by the CODEC and fed as AFR to the radio handset (Figure 3.16).

The ASIC and its associated EEPROM contain all of the security parameters, including the SKs, the key to channel assignment tables and a real-time clock.

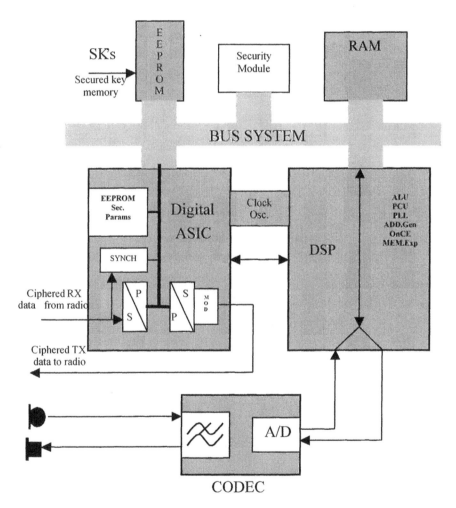

Figure 3.16 The basic architecture of a modern compact cipher module

4

Telephone Security

In the history of telecommunications, there has been no single piece of terminal equipment so widely used as the simple desktop telephone. In many parts of the world, the characteristics of being simple to operate, cheap to purchase and inexpensive to use, and the freedom to access almost any other terminal across the globe, have made the fixed line telephone a basic necessity for life. As manufacturers strive to capture their market quotas, the basic telephone has become adorned with technical innovations and services, all of which combine to produce very attractive packages that not only bestow communication quality and diversity but also position the phone as a piece of impressive furniture. The method of communication is that medium which is most comforting to mankind, i.e. the human voice, together with all of its human qualities such as voice recognition, the carrier of emotions and passions and the transmitter of all that we secretly divulge to the party at the other end of the line. In truth, our telephone secrets are divulged to anyone who has the technology and the desire to listen. The intimacy of our private lives, our commercial dealings, our political strategies and our military secrets are all 'on the line', and our message confidentiality is absolutely taken for granted.

During the Second World War, Winston Churchill was passionately addicted to his 'scrambled hot line' to President Roosevelt and unfortunately was less than cautious about the subject matter of his transatlantic discussions. Surprisingly, for one who was very well aware of the dangers of insecure telecommunications, Churchill held too much faith in the quality of his telephone encryption, and it later transpired that the Nazis were eavesdropping on his beloved hot line.

In many countries of the modern era, telephone tapping has become such a national pastime that even the general public are guarded about their telephone conversations. In other places, there are strict controls about personal privacy, which make it somewhat more difficult for an agency to monitor private telephone lines. Nevertheless, there are those who pay little heed to authorities and their ideals (Figure 4.1).

So, in such environments, if one has a need or desire to protect their telephone communications, there are several steps to take and technologies to employ. This chapter is about some of those methods. It discusses the specific threats, the technology involved in the public switched telephone network (PSTN) and satellite communications, the solutions and technology available and, finally, the different applications.

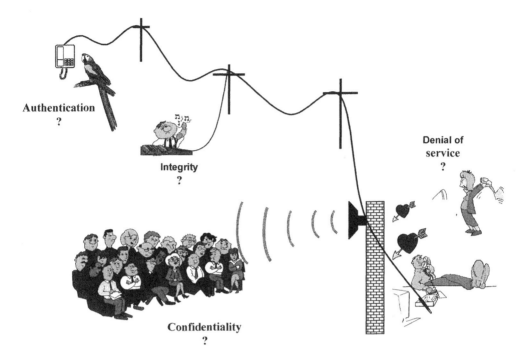

Figure 4.1   Threats to telephone security

4.1 Specific Threats to Telephone Operations

As far as the telephone is concerned, there are threats that exist because of the technology used, which leaves every user vulnerable to attack, and there are threats to secured telephone installations, i.e. cryptographic threats. The general telephone threats are:

- Bugging. Telephone lines are very easy to access, and who knows where the cables at the end of our terminal disappear to when they enter the 'hole in the wall'? Even the simple connection of an earphone, of the portable audio type, can add an ear to an analogue telephone-line conversation. So, any agency with resources to spare will have very little problem to eavesdrop.
- Electro-magnetic compatibility (EMC) or transient electromagnetic pulse emanation standard (TEMPEST) attack on radiated information from a commercial telephone, which can be considered as another form of bugging.
- Denial of service (DoS). Even a heavy storm can deprive the user of essential services, so in a state of man-made threat, the simple telephone line is even more exposed to attack. At least the DoS attack should be apparent to the user, giving them the opportunity to find another avenue for communications.
- Authentication. Is the person at the other end of the line, the one who he or she claims to be? It might be the double-glazing telesales caller, but it could very easily be a quite different party checking on their identity and location for clandestine purposes.

The threats to 'secured' telephone interactions are:

- The loss of secret keys.

- Cryptanalysis of a ciphered signal.
- Unauthorised use of secure equipment.
- Jamming of cipher communication, i.e. DoS, to cause the device to fallback to a default key or plain mode of operation.
- EMC/TEMPEST concerted attack.
- Spoofing, i.e. the loading of keys known to an 'interested party' into two or more stations and who is therefore able to monitor ciphered exchanges with ease.
- Erroneous use of plain mode by the operator.
- Poor-quality terminal equipment, which purports to give high security but does not.
- Key management problems leading to either 'blocked channels' or reliance on plain mode.
- Full automation as far as security procedures are concerned and transparent to the user.

A general conclusion would be that all of the telephone infrastructure outside a user's direct control must be considered as totally insecure and therefore in need of protection.

4.1.1 Telephone Security Requirements and Features

The applications where high-security telephone equipment is required are:

- Government bodies, e.g. ministries
- Police, drug enforcement and security groups
- International entities such as United Nations
- The military
- Royalty or heads of state operations
- VIP operations
- Paramilitary organisations
- High-level banking and commerce
- Criminal institutions

The protection measures that sophisticated, high-quality secure telephones should offer are:

- Resistance against cryptanalysis by the use of high-quality algorithms.
- Resistance against physical tampering attempting to gain access to security parameters such as keys and algorithms.
- The non-disclosure of secret key data, due to efficient key management procedures.
- Good connectivity over poor-quality telephone links, brought about by robust synchronisation techniques.
- Excellent voice-quality enhancing voice recognition.
- Operational security features such as the authentication of users, thereby supporting voice recognition as an authentication tool.
- Full EMC/TEMPEST compliancy achieved by true Red/Black separation of components.
- Flexibility to offer normal and advanced telephone operations including conference calls, fax and computer data communications, etc.
- A financially viable solution.
- Good operational procedures that are:

  Compliant with international standards

Easy to use
Comprehensive user training
After-sales customer support
As maintenance-free as possible.

4.2 Network Technologies

4.2.1 Secure Telephone Communication

A protocol in setting up a cipher telephone call is illustrated in Figure 4.2. The first stage is to set up a plain connection by dialling the number of the called station. Once the link has been established, there must be a verbal agreement between participating parties to change to cipher mode. This initial conversation, or 'PLAIN COM', is a dangerous period in the communication, and users must be very disciplined about their procedures. Much useful information can be derived by any eavesdropper catching both the initial and final stages of the conversation. As few opening verbal overtures as possible should be made before continuing the conversation in secure mode.

The switch to cipher mode is initiated by the caller in period 'A' of Figure 4.2 in the form of a dual tone multi-frequency signal (DTMF), which is received in period 'B'. There follows a 'modem training' session, during which the condition of the communications channel and modem compatibility are tested so that a suitable digital data rate can be determined.

Once the decision to go to secure mode has been made, the cipher units at each end must carry out a key agreement whereby the two machines investigate each other's key set-up and come to an agreement on which key to use for that call.

Periods 'C' and 'D' reflect the time taken for the caller and called stations to exchange details about what applications options and what keys they have available. Also there might be some examination of the form of encryption that will be used. Period 'E' is where a decision is made as to which precise key will be used for that session and 'F' the actual

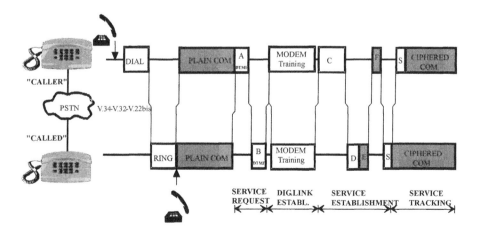

Figure 4.2   The ciphered call set up with PSTN communications

switch over to Cipher mode. 'S' is the data start signal, and thereafter, the 'ciphered comm' can take place.

The step to enter the secure mode is usually free for either party to the call to initiate, but should there be reason to switch back into plain mode, it is good practice to require both parties to make the switch command, i.e. a concurrence. Once again, verbal exchanges during this time should be brief and innocuous.

4.2.2 INMARSAT Communications

In this day and age, it is not possible to talk about telephone communications without considering satellite systems. Of the systems, the International Maritime Satellite Organisation (INMARSAT) is arguably one of the most interesting as a telephone communications carrier and became especially so when it introduced the Global Area Network (GAN) service on the M4 system. Its original mandate in 1979 was to provide marine communications by means of a satellite system, to allow ship management and provide for distress warnings. Since that time, INMARSAT has spread its wings to cater for mobile, land and aviation applications. The planned addition of the INMARSAT I-4 satellites that are due to come on line in the year 2004 will further support the Broadband Global Area Network, or B-GAN as it has been christened. The new generation will increase the present 64-kbps data rate up to 432 kbps.

Today, the INMARSAT system allows telephone voice and data communications to reach almost anywhere in the world. It operates four INMARSAT satellites, plus one spare, all in geo-stationary orbit, some 22,000 miles (36,000 km) above the Earth, with each satellite's footprint overlapping to give almost complete coverage of the Earth's surface (98%). Servicing the satellites are 24 land earth stations (LES) and an ever-increasing number of subscribers. The availability, at increasingly attractive prices, of the new generation of mobile earth stations (MES) has attracted many customers, who find a very flexible system of several standards, i.e. A, Aero, B, C, D, M and mini m, providing the following services.

- Analogue telephone voice transmission with a 3.1-kHz bandwidth.
- Data speed of 64 kbps
- Facsimile transmission
- Internet operations
- Telex transmission
- Multimedia
- LAN/WAN communications
- Video transmission

The *MES* services offer ISDN functionality carried at data rates of 64 kbps with a variety of interfaces, including ISDN, RS 232, and RJ11, and with Infra Red and USB ports. The terminals are easy to assemble and easy to operate, with typical weights of between 3 and 5 kg, and they are readily transportable to the most demanding of Earth's environments. This makes the system ideal for military, security and journalistic operations.

The procedure in establishing an INMARSAT communications is as shown in Figure 4.3. The MES sends a request on a predefined calling frequency, to the nearest Land Earth Station (LES) via the 'local satellite'. The LES responds with the necessary information on a signalling channel for the MES and allocates the channel to be used for the link communications.

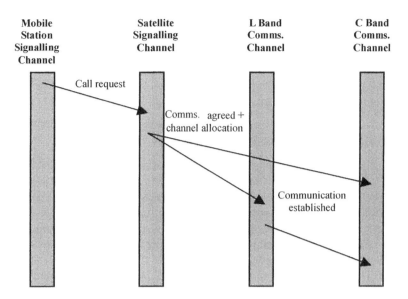

Figure 4.3   Using the INMARSAT call procedure

The LES also forms the gateway for calls to be linked between the MES and to the called destination whether it be at a fixed or mobile location via the public land based systems such as the PSTN.

To make a call from one MES to another, the caller simply dials the number of the other MES. Each MES can have one or more numbers per service type. If the user wants to set up a 64-kbps data connection, then they dial the other unit's 64-kbps data service number. Similarly, if there is a requirement to make a voice 4.8-kbps call, the caller dials the remote MES's 4.8-kbps number. The INMARSAT phone numbers have a country code of +87 followed by a digit identifying the satellite (1, 2, 3 or 4).

Many service providers, like the Norwegian Telenor, automatically detect the satellite if the caller enters 0 instead, followed by a nine-digit telephone number. If the caller wishes to dial from an MES to any other location in the world, they simply dial the international number of that location.

As far as security is concerned, just as with the normal telephone-system operation, the entire link follows unsecured paths and is therefore vulnerable to all the usual attacks. Therefore, encryption is essential to those with secrets to keep secret. A common offering made by most, if not all, MES producers is the inclusion of interfacing that gives compatibility with secure telephone unit (STU) III/IIB (see Section 4.3.1) secure voice encryption (Figure 4.4).

One advantage of the STU system is that when it comes to secure transmission over INMARSAT via mobile GAN terminals, it interfaces easily with the terminal. The output from its V32 modem is a digital signal corresponding to the digital interface of the terminal. In contrast, the normal analogue telephone output with a bandwidth of 300–3300 Hz is passed through a Vocoder (Figure 4.5).

Many of the INMARSAT services have a very narrow bandwidth available for their applications, and to accommodate terminal requirements, voice compression algorithms are utilised. This poses a problem for those services such as telephone applications that are

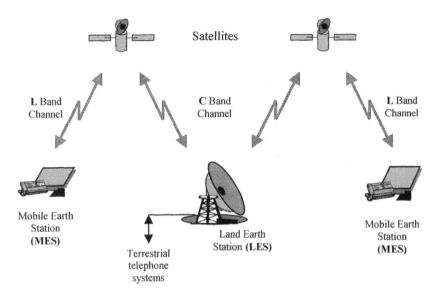

Figure 4.4   Telephone communication links over INMARSAT

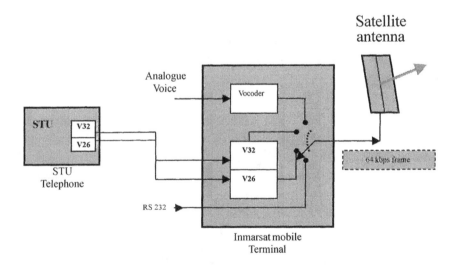

Figure 4.5   STU and analogue telephone connections to a mobile satellite terminal (MES)

encrypted, and in such cases, these compression techniques must be bypassed. There are generally two types of compression techniques: those that allow for exact recovery (data compression) and those that allow for an approximate recovery (voice, image compression). While data compression of encrypted data is possible, yet not effective, voice or image compression of encrypted data will make decryption impossible.

MES devices such as the World Phone and World Communicator from NERA SatCom, (Figure 4.6) implement software mechanisms, which allow encrypted data at 2.4 kbps to pass

Figure 4.6   The NERA World Communicator (Courtesy of Nera Telecommunications Ltd)

through with transparency. This solution must be implemented at both the MES end and the LES end of the link. One of the main features offered by the Nera World Communicator is the 64-kbps ISDN line, and using the 3.1-kHz ISDN feature, the STU III can be connected directly through the MES as can voice and data signals operating at 9.6 kHz. The situation for phone and fax communications digital or analogue over the INMARSAT is shown in Figure 4.7.

The Nera, WorldCommunicator interfaces include:

- ISDN
- RS-232

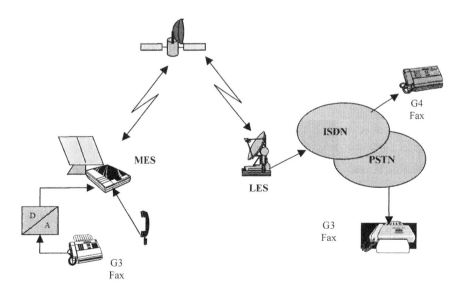

Figure 4.7   Digital and analogue communications over INMARSAT

- USB
- Built-in DECT base station for wireless handsets

The terminal uses any type of ISDN device or devices with an ISDN connection. It has a built-in point-to-point protocol (PPP) modem that converts the asynchronous signal on the RS-232 or on the USB, to a 64-kbps bit stream (ISDN UDI), facilitating connection of a computer to the terminal. The Nera INMARSAT GAN unit has a built-in DECT base station: this is small and light, and a compact unit is available. It has IP grade 55 and is designed rather to fall apart and be put together again than to break, if handled too roughly. It is designed to be a satellite data modem with voice capacity.

Whichever system is used, telephone conversation becomes insecure from the moment it leaves the handset microphone at the transmitter, until the faithful reproduction is emitted by the receiver's earphone. If one is tempted to rely upon the various and complex technologies involved in communicating telephone messages around the world, for their security, then one should think again. There are those proponents who point to the communications technologies of time division multiple access (TDMA), frequency division multiple access (FDMA), signal digitisation and quadrature modulation as being elements with protective characteristics. To some extent, such comments are reasonably well founded, as the opportunity to pluck phone signals from the line or air is not as easy as it was for the simple analogue systems from a generation ago. However, communications receivers are designed to do just that, and the technology can be modified relatively easily to build customized, eavesdropping equipment by motivated parties. Another source of monitoring technology is the test equipment that is produced by systems manufacturers. The purpose of this equipment is primarily to maintain the telephone system, and amongst its many tests is the ability to capture any telephone call, at any point in the route that it takes. Government agencies around the world are blessed with the financial resources to install the latest state-of-the-art test or monitoring equipment and certainly do so. One such system of monitoring GSM telephone calls is described in Chapter 5. It is the same situation for PSTN telephones, so when making any call, be it local, national, intercontinental, satellite, line or optical, the best policy is to adopt the philosophy that somebody *is* listening. Bearing in mind the often torturous route that a telephone call might take, end-to-end encryption can be seen as the only secure method of ensuring message confidentiality and integrity. The remainder of this chapter discusses ways on how these and message authenticity may be achieved.

4.3 Telephone Security Solutions

There are any number of 'secure telephones' available on the market, and, as one would expect from a telephone, most have many features in common. This chapter, however, approaches the subject by drawing comparisons between two scenarios of ciphered telephone operations. One is a US government-sponsored solution, the STU, and the other represents a proprietary solution that is more or less off the shelf, available from a wide variety of manufacturers. Both systems are able to operate in the normal manner, i.e. desktop to desktop, in plain and ciphered modes, using the PSTN, and both can be applied to satellite communications using, for example, the INMARSAT system as described previously.

The operations between GSM and desktop telephones are dealt with in Chapter 5 and so will not be repeated here.

4.3.1 STU III/IIB

The 'III' in STU III indicates the third generation of secure telephone unit. As the name suggests, it is a secure telephone manufactured by AT&T, GE and Motorola but under the control of the NSA, the National Security Agency of the US. The first version was released in 1970, followed by version II five years later, with the subsequent version III in 1987.

It seems that the main motivation behind the product was the fear that defence secrets were being lost to eavesdroppers listening in on conversations between US defence organisations and their contractors. Hence, the deployment of the NSA sponsored STU. It was considered in the USA that a solution to the leakage of sensitive data at lower levels of concern was providing fragments of data from which a big picture of the American defence security status and policy could be extrapolated. Implementing security like STU and distributing it widely throughout sensitive industries was thought to be a big step in the right direction and, with an estimated 200,000 units allocated to interested parties, gives an idea of the concern and effort to moderate the leakage.

The STU III has four categories of encryption:

1. Machines that are restricted for use by the US government and military when transmitting classified information between themselves and defence contractors.
2. Machines that are restricted for use by the US government for sensitive, but unclassified, communications.
3. Machines that are exported to and operated by US corporations operating overseas.
4. Machines that can be exported to any organisation except those belonging to the US black list of prohibited organisations or countries. The actual export is dependent upon the issuance of an export license.

STU IIB has only one category, and that is for NATO governments' operation, with the additions of Australia and New Zealand.

Whilst four categories of STU III security are publicised, it seems quite likely that this might not be the complete picture, and it seems sensible to assume that there is a probable fifth category, if not more. Depending upon the numbers distributed and the key management organisation of cipher units, one would expect to see an ultimate category for 'Top Brass' communications within the US government agencies. If indeed this is the case, we can be certain that the fact will not be made public in the foreseeable future.

The STU system is aimed at essentially forming something of a closed network or series of closed nets. For those operating within the system, it would be highly acceptable, and its implementation has been welcomed by many contractors to the US government. Participants have become increasingly aware that millions of dollars are being lost to illegal, corporate eavesdroppers. For the rest of the world, the fact that the NSA sponsors the STU is a message of sufficient alarm, to direct those searching for a more independent secure telephone communications system, to look for an alternative solution.

STU III has earned a high reputation for hardware quality, particularly with its strict approach to Red/Black separation. It is constructed in two distinct parts, the 'Black' telephone section taking the upper part and the Red security structure being mounted underneath with full metallic shielding in between.

Being very much, a top-secret device, the system key management can be assessed only speculatively. Whilst the NSA is undoubtedly the 'big brother' with his hands on the controls, it seems probable that there are a number of key management stations under the central body. This conclusion is arrived at when considering the enormous task of administering key generation, distribution and agreement, to a vast number of machines that are spread widely throughout the United States and beyond.

The STU III system itself consists of two basic components. The first is the actual desktop, telephone terminal, and the second is the key management system, which includes a crypto ignition key (CIK). The CIK is in fact a 64-kb electrically erasable programmable read only memory (EEPROM), which is used as a portable access token and key carrier or key storage device. Hence, it is often referred to as the 'KSD'. This device is programmed for clients by the NSA using the electronic key management system (EKMS) and contains the client-specific parameters that may include a seed key and/or an operational key. When inserted into the STU, they initialise the cipher process and, in doing so, convert the STU from a plain telephone into a secure, cipher unit. The controlling agency that produces the CIKs distributes them to STU III users, who must call the management centre to validate both the CIK and the STU unit with which it will be used. A typical scenario would be that perhaps eight users would be allowed access to one particular telephone station but that each of the users might also have access to a number of other stations using their personal CIK. The encryption process, once plain communications have been established, is to enter the CIK into its receptacle and switch the STU into cipher mode. The insertion of the CIK will probably initiate the release of the communication key to be used for encrypting the voice session, by using the CIK contents to decipher the communication key thereby releasing it from its storage key and making it available as a session key. It is reported that once the call has been terminated and the CIK removed, the communication key is reciphered with a different storage key and held securely until the next user accesses the unit. If this supposition is correct, it is likely that the actual data carried by the CIK are more like a certification parameter to release the communication key than an actual key encryption key, as might be initially supposed.

For successful encryption and decryption of a voice conversation across a link, there must be a key agreement process, and, given the nature of the beast in question, it is almost certainly of an asymmetric type. The typical procedure would be that when switching to cipher mode, the remote, called party would transmit their public key to the calling station, to cipher and hence protect the transport of the caller's session key to the called station. The called unit would decipher the caller's key by using their private key, and the resulting key would be the replica of that communication key (or session key) at the calling telephone. Having performed the key agreement, the conversation would be protected by a symmetrical process.

It seems most unlikely that the STU system uses DES for the encryption process as one could not imagine the NSA condoning the use of a considerably aged and weak, 56-bit key to protect its own government's communications. The arrival of AES might well see its implementation into the STU system.

4.3.2 The Alternative Telephone Security

Of course, STU is neither available to, nor a desirable solution for, everyone. The alternative secure telephone would envelop the essential qualities described in Section 4.1.2 and of the STU but add the independence of customer specific or proprietary encryption. This means the adoption of strong, customisable, multi-algorithm solutions in order to achieve the best security possible.

The algorithm itself depends very largely upon the application, the method of implementation (software or hardware) and the required processing speed. After all, it is the algorithm that provides the security of telephone messages as they fly through the unprotected media of line and ether, and is therefore one of the most important elements in a security system. Whilst there are many algorithms around, many purporting to be unbreakable, if not independent, for the client requiring high-quality encryption, proprietary algorithms that offer customisation are the only answer. If the algorithm is considered to be strong, security lies in the keys used, and therefore, standardized algorithms are definitely an option. However, some users still appreciate having their own algorithm.

It is much more reassuring for a client to be able to modify, within certain controls, an algorithm that is not promoted by a third party. Manufacturers not able to market such devices, for whatever reason, are at a major disadvantage in attracting the discerning customer who will be looking for the following features for their telephone security:

- A multi-algorithm designed system
- Customer-specific versions available
- Customer freedom for algorithm adaptation
- No export restriction imposed by the manufacturer's ruling government
- No third-party approval of the algorithm
- Physical protection against copying the algorithm

A multi-algorithm ciphering system is one that does not rely upon a single algorithm for its security. It may well consist of the following:

- Algorithm A, for ciphering 'Home network' communications
- Algorithm B, for ciphering 'Allied network' communications
- Algorithm C, for ciphering 'World network' communications
- Algorithm D, for securing key management procedures, i.e. the encryption of key distribution devices, such as chip cards or fill guns etc.
- Algorithm E for the encryption of 'down-line' or OTA 'over the air' key distribution from a management centre
- Algorithm F, an internal processor algorithm to protect parameters against readout or modification

The advantages of adapting a multi-algorithm strategy are that the function-specific algorithms are more efficient as they only address one security process. An attacker would probably need to break more than one algorithm to to be able to read the system messages. In addition, it leaves more scope for the customisation of security parameters. Multi-algorithm systems are also often necessary to cope with the different technical aspects such as operating mode and speed that are required by different applications. A graphical representation of the multi-algorithm components of a telephone-ciphering device is shown in Figure 4.8.

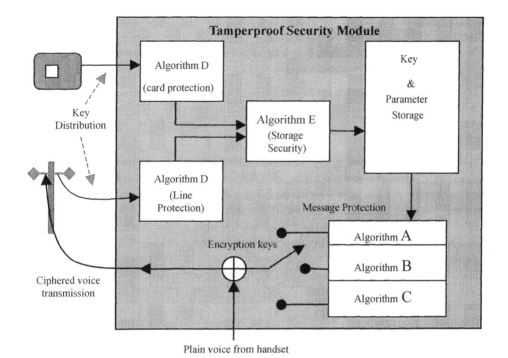

Figure 4.8   The implementation of multi-algorithms

Needless to say, the algorithm must be stored in each encryption device, and great lengths are taken to protect sensitive data such as algorithms and keys against readout or modification. Tamper-resistance/proofing techniques can protect against all but the most sophisticated high-tech attacks. Tamper resistance can be implemented physically and, for even greater security, can be supplemented by the use of logical tamper resistance, which can be defined as the resistance to gain access by electronic means. The inclusion of all sensitive encryption data into a single ASIC device that is internally protected by storage encryption will render all attempts to access the data infeasible. The process is based upon a unique key resident inside the processor that is used to encrypt user-defined data stored in an EEPROM or random access memory (RAM).

The secure telephone must have a fail-safe capability to protect against the possibility of the cipher unit falling back into plain mode, should a fault condition occur. Under such circumstances, either it must be impossible to operate in an insecure manner or at least a clear alarm must be given to the user to identify the condition. The quality of telephone lines in many parts of the world is not what might be expected in the developed world, and so in order to maintain connection links, the secured phone must exhibit robust synchronisation of the cipher units. A modem with automatic fallback from typically a maximum data rate of 9.6 kb/s, used under ideal conditions, down to 4.8 kb/s or even 2.4 kb/s in the worst case is a standard requirement.

As the security of a telephone network depends essentially upon its secret keys, as does any ciphering communications equipment, the key management is an essential factor to consider when implementing security. The judicious client will be keen to assess what support to their network security is available in this aspect.

Controlling access to a secure telephone is also a subject for contemplation and careful management. Various strategies of access that can be employed include:

- Open access, with freedom to choose plain or cipher mode, to all users
- Access to all secure telephone stations but only for users with rights
- Access to selected stations allowed only to selected personnel but with common access to all other stations

This issue will be dealt with in detail later on in Section 4.4. For the moment, it is at least apparent that when controlling access to a secured telephone network, it can become a complex matter and one that cannot be efficiently controlled simply through the use of passwords. Access tokens in the form of dongles or chip cards find an ideal application here, as they can be used not only for access but also as the carriers of personal secret keys. These tools bestow flexibility upon the system and provide a facility often demanded by senior, mobile personnel. Hence a provision for such applications would involve the inclusion of an access token reader within the hardware and would definitely be an attraction for the consumer organisation. The STU has its own access token and key carrier in the form of the CIK carrier.

The quality of present-day telephones is such that voice recognition can be sufficient to give authentication to the users, but this is not always the case. To supplement this inherent and fundamental feature, further identification of the remote party can be implemented. Authentication can be enforced by:

- The exchange of telephone numbers
- The exchange of secret key signatures
- The exchange of personal IDs

The first of these options is far from a satisfactory solution as it is quite simple to program a telephone in order that it transmits a bogus number during any exchange. Having said that, however, the programming of a secure telephone with parameters including its ID should be protected by access restrictions imposed by the security manager. The exchange of secret key signatures can at least give some measure of authentication support by inferring that the user and/or station, if keys are resident in the unit, are part of the same network. Exchanging key signatures by the Diffie/Hellman exchange provides the opportunity to authenticate stations/ users with a certifying authorities certificate, and the use of signatures guards against the 'man in the middle' problem that dogs the Diffie/Hellman process. Authentication can be taken a step further by using security parameters, particularly identifying data that might be held in an access token, in the handshake exchange between units. The authentication becomes stronger when such a policy is adopted as it then relies upon a package of elements rather than a single entity, i.e. user password.

Figure 4.9 shows the general idea of user authentication, although the actual procedure can be 'very application-specific'. User authentication at the cipher unit typically uses the philo- sophy of 'something known' with 'something owned', and here, the chip card is used as the latter, and the known component is the user's PIN or password. The user enters their PIN at the unit's keyboard, and the cipher unit authenticates the user after verification of the card

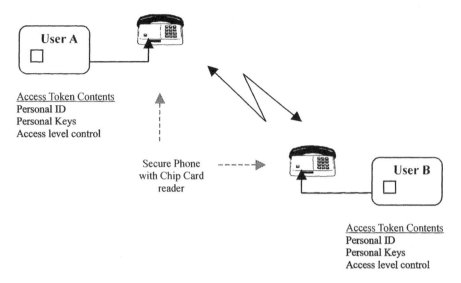

Access Token Contents
Personal ID
Personal Keys
Access level control

Secure Phone
with Chip Card
reader

Access Token Contents
Personal ID
Personal Keys
Access level control

Figure 4.9   User authentication by exchange of security data

content and also possibly with the cipher unit itself. The cipher unit may be pre-programmed with a user access table and facilities, depending upon the application.

4.3.3 Hardware Security Features

Sometimes, too great an emphasis is placed on the aesthetic appearance of a telephone when deciding on which to purchase. As inferred earlier, the telephone enjoys a special niche in office and home furnishing, as well as being a tool for communications. Modern design and materials, especially the use of plastics in manufacture, give producers a large amount of scope in meeting the demands of customers. However, the important thing to remember when put in the decision-making position is that looks are not everything and that cosmetic appearance should not be at the expense of security. This is particularly the case when assessing a machine's ability to provide protection against EMC/TEMPEST related compromising radiation. Ironically, the older and perhaps cheaper telephones are not really regarded as a serious compromising emanation hazard as their simple functionality often avoids the problem. Rather, the analogue-to-digital converters with fast edge signals and use of digital signal processors (DSPs), perhaps for echo cancellation, of the more sophisticated telephones, create greater problems. In these cases, there is little alternative to the metallic shielding of the telephone and careful attention to circuit-board design and prudent component assembly in overcoming this problem. It is unlikely that the manufacturer of 'off the shelf' telephones is going to spend much attention in compromising radiation and Red/Black separation characteristics other than those necessary to meet the regulations set by the national or CITT, which includes some EMC regulation. The cost of secure protected telephones, built to much more exacting specifications, would probably prohibit their purchase by the general public. Therefore, it is most unlikely that such a device could be modified to give effective protection when included in a secure telephone network. The purchase of specific, purpose-built hardware is the only sensible option and one that is, by the laws of supply and demand, likely to mean an expensive outlay. There is an in-depth discussion on the subjects of EMC/ TEMPEST in Chapter 1.

4.3.4 Telephone Security Architecture and Functions

This is best illustrated by Figure 4.10, which shows the basic features of the security architecture.

The plain voice signal path follows the normal telephone processing, bypassing the modem and vocoder, i.e. there is no processing whatsoever as far as security is concerned. Indeed, the plain signal path and the secure operation are the main focus for Red/Black separation.

Red Boards carrying secure components would include:

- Voice-coder module
- Main board containing the main and cipher processors
- Smart-card reader modules except where the data are encrypted
- Entry devices that allow the key inputs to be made in plain, e.g. keyboards

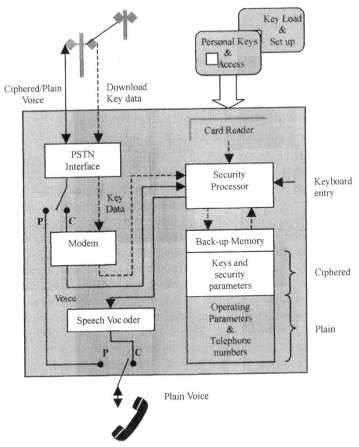

Figure 4.10   Secure telephone architecture

Black Boards carrying non-secure components would include:

* Modem
* Modem controller
* Telephone interface

Key data can be entered into the cipher unit, typically in three ways:

1. Manually through the telephone keyboard.
2. By chip card, fill gun or dongle.
3. By 'down line loading' (DLL).

Manual key inputs through the keyboard, is a very simple and usually basic equipment feature, but as this method is usually a plain entry, then if the keyboard is of the Black classification, a security problem may exist.

In the chip-card process, the card carrying the keys and parameters is entered into the reader. The data stored on the card must be protected by encryption with a transport key, and therefore, a copy of that key must already exist in the telephone cipher unit. After the transfer from the card, and once they are deciphered, the keys and parameters are loaded into the secure memory where they should be ciphered once again with a storage key and remain ciphered until use.

In the same manner, the DLL of security data is also protected by a transport key. The DLL contents are enciphered on the 'packets' initial programming at the management centre and then deciphered as the packet reaches the security module inside the telephone. The sensitive values are passed through to the security processor, which once more encrypts the data with the storage key storing the product in the secure memory until required.

It is probable that a ciphered telephone will have options including fax and data transmission, and if this were the case, another input/output interface, e.g. V24, would be added to the unit. Such an interface is not shown in Figure 4.10 for reasons of clarity, but it would normally be located between the vocoder and the security processor.

As can be seen in Figure 4.10, the vocoder is bypassed when the unit is operated in plain mode, but its inclusion in cipher mode is to encode the voice signal within a narrower bandwidth than would be possible by just using a D/A converter. This implies a better voice quality, given the bandwidth of the telephone channel.

4.4 Key and Access Management

Depending upon the operating access programmed by the security manager, an individual user might be required to use their card or dongle, not merely as an access token, but also as the carrier of their personal secret keys. These are the keys that the user will use to cipher their calls. Of course, this gives the user the flexibility of using their keys at any secure telephone station within the same group. All keys carried in the token will be ciphered, at the source, i.e. the management centre, by a key transport key (KTK) to prevent the possibility of their being read out from the token in plain.

The personal key is read out from the card/dongle, then deciphered by the transport key and loaded into the security processor where it is temporarily located and used for that ciphered session. Once the call has been terminated, the user withdraws their card and, in doing so, removes their keys from the terminal thereby preventing their being used by another caller thus enhancing the user's own authentication status. This situation is ideal for senior personnel or where the use of the secured telephone network is highly restricted.

The key agreement/selection process should be capable of resolving a key priority decision. This is essential in cases where two or more stations have more than one common key to choose from. In such a situation, it is important to select the key according to the correct criteria; otherwise, it is possible that the same key would be used under all circumstances. Various criteria can be used for the agreement process, including the security level of the key, age, period of validity, etc., all of which could be incorporated into the key's ID structure.

In a larger network, perhaps with little hierarchy and with a high usage by large numbers of staff, the policy of each user carrying their own keys in a portable medium might be difficult to manage. In this case, it would probably be safer to have keys permanently stored in the telephone units themselves and not rely on individuals to be disciplined about their communication activities. This mode of operation really depends upon the structure of the network, its size and the ease with which key distribution can be carried out. There follow some suggestions on how network management might be carried out.

4.4.1 The Complete Key System

In high-security systems, one cannot expect communications to be protected by the implementation of a single key function, i.e. restricted to just securing the message data. In practice, this fundamental feature must be supported by a whole package of keys, each performing a different task and representing another hurdle to be cleared by any would-be assailant. A typical key structure might include the following:

- Secret key (SK), or communication key
- Broadcast key (SK-B)
- Key transport key (KTK)
- Card encryption key (CEK)
- Management database key (MK)
- Tamper resistance key (TRK) or storage key
- Key storage key (KSK)
- Various default keys (xxDK)

4.4.1.1 Secret Key (SK)

This is the key that is used to cipher the actual voice or data communication. It may be used to give direct encryption of the message or used in a key agreement feature to generate a session key. Its typical key length will be 128 bits made up of 32 alphanumeric or hexadecimal characters. Hence, the typical key variety would be 16^{32} or $10^{38.}$ A cipher unit may contain up to about 100 SKs and can be used to define the network topology, as illustrated in Figure 4.11.

The SKs would be entered into the cipher unit by manual, key card, token or dongle and by DLL. They would be secured in the cipher unit memory by the KSK.

4.4.1.2 Broadcast Key (SK-B)

It is common practice these days to engage in conference calling, and in the case of a secured

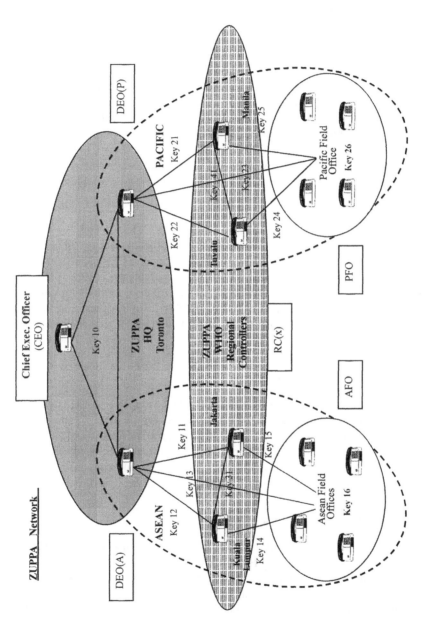

Figure 4.11 The ZUPPA network topology

conference or broadcast call, common SKs must be available to all conference participants. A number of SK-Bs should be generated and distributed amongst users in a sub-hierarchy arrangement. A key agreement protocol using a priority list of keys might be the easiest and most efficient method of achieving agreement and synchronisation with a group call.

4.4.1.3 Key Transport Key (KTK)

The KTK is used in SK downloading over the communication link from a management centre and is used to encrypt the SK prior to its download and decrypt the SK at the telephone cipher unit. Hence, the KTK is used to form a logical 'secure channel' for the SK distribution. As far as the key manager is concerned, this is the most important key of their pack, as the confidentiality of whole SK distribution depends upon its secrecy. A well-structured network will use multiple KTKs with, ideally, a different KTK for each secure link. Therefore, the compromise of a single KTK will only affect the security of that particular link, whereas if only one KTK were used for the whole network, the entire network would be compromised.

4.4.1.4 Card Encryption Key (CEK)

It would be possible to use the KTK for this purpose, but a higher security can be achieved by splitting the SK distribution protection between the KTK and the CEK. As its name suggests, the CEK can be used to encrypt the SK, and any other parameters such as passwords, for transport to their destination units. The card or dongle distribution method might well be used to carry other sensitive key parameters such as the CuK, and so it makes sense that that key should be protected by a different transport key other than the KTK as it performs a different purpose.

4.4.1.5 Management Database Key (MK)

The MK is resident in the networks key management centre and is used to encrypt the network parameters that are stored in the management centre memory. As the management centre contains all of the network data, it is a prime target for attack, and the loss of the database, even though it may be backed up, is a major threat. Ideally, the centre will be located in a secure area with a very restrictive-access strategy enforced, but in any case, the sensitive data should only be stored on removable hard disks encrypted by the MK.

4.4.1.6 Tamper Resistance Key (TRK)

Tamper proofing is usually thought of as being solely a physical feature to protect critical electronic circuit hardware from attack. However, as mentioned above, this is not the whole story, as a high-security module requires logical tamper-proofing as well as physical. The idea is that sensitive parameters such as algorithms and keys should be resident within a single chip and should never be visible outside that device in plain mode. The TRK is a unique key that is stored in a physically tamper-proofed chip. No copy exists except that inside the security module to which it is specific. This key is used to encrypt the user-defined parameters that reside within the internal non-volatile memory of either the security module, or external to it, and hence protecting it against logical tampering. If not so protected, entities such as keys would be vulnerable to modification and therefore to 'biasing' the subsequent encryption process, in favour of the invading party.

4.4.1.7 Key Storage Key (KSK)

When keys, i.e. the SKs, are stored in an encryption device, which may or may not be within a tamper-proof or -resistant module, it is common practice to protect these keys from illegal access or modification. The KSK may be distributed by a chip card as part of a security parameter package and thus protected by the CEK.

4.4.1.8 Default Keys (xxDK)

There must exist a foundation on which to start building a secure system, and the use of default keys provides just such a base. Their purpose is twofold. In the first instance, they are certainly not intended as security devices in that they are used to encrypt messages, as a privately generated SK would. Being default values means that they are quite likely common amongst the 'machine breed' or at least customer type and should therefore be regarded as being weak. However, they do serve the purpose whereby if the keys that have been programmed into a cipher unit have been lost for one reason or another, ciphered communications can still take place using default keys albeit weaker than should be the case. Anything but plain! This has often been the case in the past when the consequences of transmitting in plain have been severe. Default keys are also useful in maintenance and in the initial set-up of networks. With default keys permanently installed, a key manager can concentrate on setting up the physical aspects of a network without worrying initially about the security topology. Once the physical things have been addressed successfully, the manager can use the default keys to operate in a basic cipher mode and at least establish the network before securing it with custom generated keys. If problems persist, then a complete reset can be easily carried out to reset all stations to the default status and start again from this reference position. It is a very useful asset, especially when a network is widely distributed geographically, and troubleshooting has to be done remotely. In any case, the use of default keys as fallback keys must only be a temporary event and carry no sensitive data during that period when they are used.

Section 4.4.2, describing the 'ZUPPA' network, illustrates how some of these keys can be put to use along with some thoughts on access rights.

4.4.2 The 'ZUPPA' Network

The task presented before the security management is to design and implement a secure telephone network for the ZUPPA organisation, whose headquarters we assume to be in Toronto, Canada. It is also the normal seat of its Chief Executive Officer (CEO) and their Deputies DEO (A) and DEO (P), although they are expected to travel within the organisation. The DEOs are officers representing their respective, ASEAN and Pacific regions at the HQ. Each region is also represented by Regional Controllers (RC (x)), who are located at the desks identified in Figure 4.11 and who are in control of day-to-day matters around the field offices (AFO) and (PFO).

4.4.2.1 Security Policy Definition

The brief given to the design team consisting of a Communications Officer, a Security Manager and a Key Manager, is to provide for:

- High security and flexibility for the senior players.
- Transparency of use for the RCs whose own task is to administer the field operations of their health workers, i.e. FOs. They are not to be burdened with security operations.
- Segmented regional links, ASEAN and PACIFIC with cross-communications only at the highest levels in the management hierarchy. The segmentation is sensible considering the possible differences in political and ethnic attitudes between regions.
- Ease of operation for the field personnel, who, being locally recruited staff, will probably be neither experienced nor very interested in security control. They will only have the option to use cipher mode operation within the network.

So, this is the policy developed by the security team and their client.

The communications manager's prime objective is to establish and maintain the physical communication links throughout the network. This will involve the technical aspects of telephone networks, PSTN, ISDN, radio and satellite links, etc. For the security manager, these are of less concern.

The Key Manager's role will be to carry out the day-to-day generation, distribution and changes of keys and monitor access rights throughout the network. The Key Manager may well share the role of developing the security policy with the management team but will not have the rights to modify it.

The Security Manager's role is the formulation of the policy for ZUPPA's security and to oversee the implementation, thus including:

- Link protection
- Key hierarchy
- User access rights
- Key Manager's and Communication's Manager's access rights

Once the policy has been fixed and the network management team selected, the implementation steps should follow the following steps:

1. How to establish individual access rights.
2. How to grant the flexibility or impose restriction of individuals.
3. How to implement the highest security.
4. How to bestow transparency for the users.
5. How to grant ease of operation for semi skilled field workers.

4.5 Network Implementation

Each of these issues is better dealt with according to the layer structure implied in Figure 4.12 and can be summarised in Table 4.1

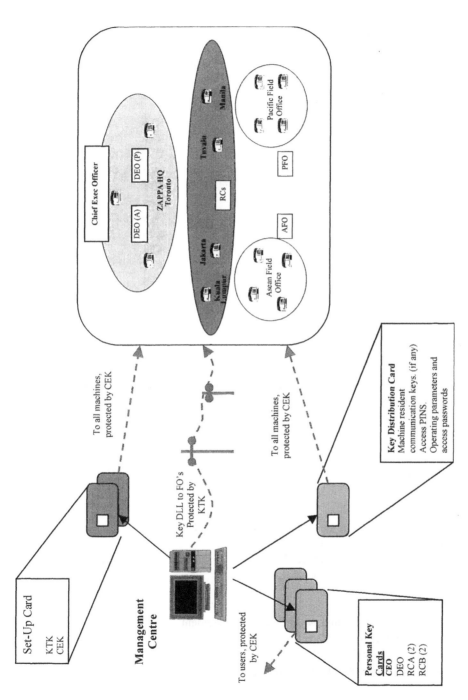

Figure 4.12 Key management and distribution for ZUPPA Net

Table 4.1 Network parameters

User	Access rights (location)	Card content	Access by (i) PIN and card; (ii) pass/phone	Phone contents and parameters	Operation mode P, C or free
CEO	All stations	Key 10	i	Personal use only	Free
DEO (A)	All stations	Key 10	i	KTK KTK Use: DEO (A + P), CEO	Free
DEO (P)	All stations	Key 10	i	KTK Use: DEO (A + P), CEO	Free
RCA (K)	RCA (K + J)	12, 14, 31	i	KTK	Ciphered
RCA (J)	RCA (K + J)	11, 15, 31	i	KTK	Ciphered
RCP (T)	RCA (T + M)	22, 24, 41	i	KTK	Ciphered
RCP (M)	RCA (T + M)	21, 25, 41	i	KTK	Ciphered
AFO	AFO only	No card	ii	KTK 13, 14, 15, 16	Ciphered
PFO	PFO only	No card	ii	KTK 23, 24, 25, 26	Ciphered
Definition[a]	a	b	c	d	e

[a] Column definitions: a: Who can use which telephone? b: Which communication keys are carried on a card? c: Access to phone by card with PIN or password alone? d: Parameters residing in a particular phone? e: Fixed mode of operation, or free for user to chose?

User parameters:

CEO and DEOs:

- Can use any station to communicate with DEOs and HQ
- Carry their personal keys on a card
- Access by their own PIN with a card
- Exercise freedom to chose mode of operation

RCs

- Can use any RC or FO station within their own segment
- Carry all personal keys on a card
- Access by own PIN with a card
- Allow ciphered calls only on the network phones

Field offices (FOs)

- Can only use xFO phones
- All keys are fixed in phone and pre-allocated to numbers dialled
- Access can only be gained by own password input
- Ciphered calls only

Note: The further down the hierarchy, the weaker will be the operating discipline. Therefore, at the FO level where the operators can be expected to be the least trustworthy, they will have almost no freedom to choose how they will communicate. Being forced to communicate in cipher mode only (also the same for RCs) is a deterrent against the use of phones for personal calls and snarling up the network with trivial calls. Passwords would normally be expected to restrict the use of phones to the organisation's personnel, but lax security at lower levels such as in FOs will mean that passwords may well become common knowledge to other residents, and 'use abuse' may become a distinct possibility. A fixed cipher mode would limit this abuse but not overcome the user authentication problems introduced by it.

It would be dangerous to ignore the idea that communications within the FOs are of a high risk, and for best security, a fully meshed key system for the FOs would give better security. However, bearing in mind the remote locations, the unreliable users and the probable key distribution problems, it might be best to rely on a single key sub system. Therefore, for internal communications, AFO will use Key 16 and PFO Key 26. The use of a single key for the FO systems is not ideal but can be compensated for, to some extent, by carrying out frequent key changes. The FO user will have no access to any function other than to use the phone keypad for dialling the called station. All key agreement is automatic.

All stations must have a CEK, wherever a card is used, to decipher the card contents, i.e. the user's secret keys.

All machines will have the manufacturer's default keys available for use in the case of an emergency clear or factory reset operation being carried out. However, useful as this function may be, it does incur security risks, and it is therefore important that warnings be displayed to the users whenever a default key is used. It is also a function that should only be enabled by the security manager after some deliberation about the possibilities and consequences of its misuse.

4.6 Key Distribution

It is worthless having a complicated network security topology if it cannot be maintained by key distribution and key changes. What looks fine on paper is very often impossible to implement or control, and therefore, with such a widely spread net, a package of key management tools should be employed.

The location of the key management centre (KMC) is in question, but we can assume that, initially, it would be located at the HQ. Based on this, key distribution to the HQ-based staff is relatively simple, and the generation and distribution of cards to the CEO and DEO can be achieved weekly with few problems. Changing Key 10 every week is to be recommended, as it will be used to protect highly sensitive data, and, as its distribution is limited to local application, it can be changed often without fear of affecting the rest of the network.

Distribution to regional controllers is somewhat more complex owing to their dispersed locations. A weekly change by card distribution is likely to introduce some problems, and therefore, a fortnightly or monthly change might be more feasible. If problems persist so that regular changes cannot be carried out, DLL might provide a better solution. This will almost certainly be the case for the FO stations where the efficient use of personal cards is in doubt in any case. Therefore, one can assume that key distribution by card to the machines installed in these outer regions is also suspect.

The proposed key management situation is modelled in Figure 4.12, and perhaps one of the most important features highlighted here is the split in security functions. This makes any interception of DLL or loss of card less catastrophic than it would be when that medium is carrying all keys and parameters.

The first part is the initial set up of the network machines, which is achieved by using two sets of smart cards.

- The first card to be distributed is the set-up card, the tasks of which could be further divided if necessary, that carries the basic means of protecting the communication key distribution, i.e.

  - KTK to decipher DLL contents arriving at stations
  - CEK to decipher key distribution card contents

- The second set of cards, the Key Distribution Cards, will be carrying the actual communications ciphering keys (encrypted by the CEK) to be loaded into the station machines, where these are required.
- The third feature is the possibility of DLL communication keys and other parameters. In ZUPPA Net, the FOs and RCs will be the prime targets of DLL.
- The fourth level is the distribution of personal keys and parameters to the SG, DSGs and RC(A) and RC(B) personnel.

4.7 Summary

Once the security management team get into implementation details, all manner of individual requirements and hazards will emerge. Security management becomes complex as the system grows, threats change and individual user parameters need modification. One of the easiest events to miss, yet one for great concern, is the movement of personnel, especially when it involves staff leaving or joining the organisation. In these circumstances, immediate key changes, cancellation of passwords and modification of access rights must be carried out immediately. Departing members of staff, especially disgruntled members, are a major source of threat to any organisation.

In the relative simplicity of ZUPPA Net, a multitude of questions arise, and the situation can get out of hand very easily. One can imagine that the CEO might want direct ciphered communications access to the RCs, and it is left to the reader to speculate on how this basic change might best be carried out. Communication networks are dynamic entities that need constant care and attention, and the most important advice that the author can give when setting out to secure networks as presented above, is to *start simple*! Get basic features operating successfully first, restrict as many non-essential user functions, as is possible and then let the network grow and fine-tune it, with time. To attempt to answer all questions and meet all demands posed at the outset of the venture is to invite disaster.

5

Secure GSM Systems

With the advent of digital mobile telephones came the misconception that the new generation was secure from eavesdropping. Alas, this is not the case, although the situation has improved somewhat. The occasions when one could hear another conversation taking place on an adjacent analogue channel, when that system was the normal mode of operation, are no longer the cause for concern. The analogue mobile system was certainly a security headache, but, as will be seen, the digital system, despite the wonders of its high-tech architecture, is still prone to the determined assailant. In fact, the global system for mobile communications (GSM) was designed to give the same level of security/confidentiality as the existing public switched telephone network (PSTN) network.

There is no doubt that the GSM service gives a much superior speech quality, lower terminal and operational costs, a higher level of security, international roaming and a variety of new services and facilities. To gain a better understanding of the weaknesses of the GSM system, this chapter continues by introducing the basic architecture and components of the system in question.

5.1 The Basic GSM Architecture

The GSM system is a cellular system that allows the re-use of frequencies by arranging their allocation in a fixed pattern known as clusters (see Figure 5.1) The sizes of the cells in this drawing vary according to the transmission power, ranging from 2.5 to 320 W, of the regional base stations, i.e. the base transceiver station (BTS). This means that the shape and size of cells and clusters can be tailored to suit the requirement of a particular environment. A rural cell may be several kilometres across, catering for a low number of calls, whereas the smallest cell is found within high-density areas, such as at airports where many channels will be required for simultaneous calling. The even higher density of calls may generate the requirement that a cell might only cover a single building. Such cells are called micro-cells, and they are regularly used to increase the capacity of the system in heavily populated cities. As a mobile user moves between adjacent cells, a rather complex handover procedure is carried out, and this imposes heavy operating overheads on the system, particularly where there is movement of mobile stations along motorways or highways. To cope with this problem, umbrella cells are constructed along such arterial routes. The BTSs of such cells usually have a higher transmission power than normal cells, and the handovers become a predictable routine for the switching centre.

In addition to the domestic network structure, the GSM operator must provide interconnec-

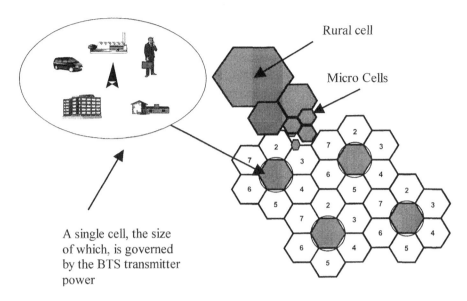

A single cell, the size
of which, is governed
by the BTS transmitter
power

Figure 5.1   The cellular structure of the GSM system

tion and operability when interfacing with the PSTN. The point of interface between the two systems is quite significant in terms of security as it is at this juncture where the GSM call leaves the somewhat protected refuge of the mobile system and is thereafter vulnerable to all of the inherent traps in the local PSTN. What might be secure on the GSM is most certainly not on the public telephone system, unless extra precautions are taken. The provider must also construct, negotiate and manage the international roaming of his customers. This also raises some interesting points regarding security.

5.1.1 System Components

The typical system components comprise:

- Mobile stations (handphone or car phone) (MS)
- Base transceiver station (BTS)
- Base station controller (BSC)
- Gateway mobile services switching centre (GMSC)
- Operations & management centre (OMC)
- Home location register (HLR)
- Visitor location register (VLR)
- Authentication centre (AuC or AC)
- Equipment identity register (EIR)
- BTS–BSC interface (Abis)
- Air interface (Um)
- GSM algorithms A3, A4, A5, A8
- Secret keys Ki & Kc

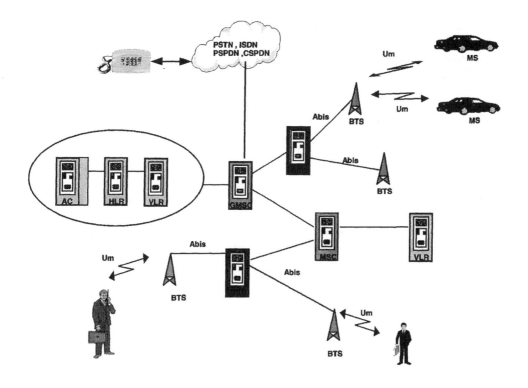

Figure 5.2   The interconnection of GSM system components

The connections between these components and their implied functions are illustrated in Figure 5.2 and described in the subsequent paragraphs.

MS: This is the personal hand phone or car phone with a transmitting power range of 0.8, 2, 5, 8 and 20 W. The actual power used is established as a result of an automatic agreement with the corresponding BTS and uses the lowest suitable power to maintain the link.

BTS: This is usually located close to the centre of a cell, and it is the output power, in the range of a few hundred metres to a few kilometres, of the BTS, that is the deciding factor for the cell size. Each BTS normally is made up of up to 16 radio frequency (RF) telephone channels.

BSC: This is the base station controller, controlling several BTSs. Depending upon the network size, they may handle from less than 10 to several hundred BTSs.

GMSC: This is the interface between the GSM cellular system and the PSTN. It controls the routing of call to and from the GSM network and carries information about the locations of the individual mobile MS.

OMC: This is the system monitoring the error messages and component status reports from the other network components. It configures the BTS and BSCs and has a controlling hand in the traffic load of these units.

HLR: The home location register contains all the individual user details for that region of the corresponding GMSC. One of the fundamental components of GSM security, the international mobile subscriber number (IMSI) is stored here, along with the user's authentication key, telephone number and billing details. It is a centre for security control and therefore of some considerable concern.

VLR: The visitor's location register plays a major role in GSM security operations. It carries the pertinent details of any visiting mobile to the region of the corresponding GMSC including the temporary, mobile subscriber identity (TIMSI), used in the authentication procedure of its visitors. The VLR also provides details of the locations of mobiles to the GSMC to facilitate the routing of calls.

AuC: The principle function of the AuC is to house the algorithms for the GSM authentication of mobile stations. It is therefore a major factor in the GSM security function and is protected from illegal access and violation

EIR: The EIR carries details, i.e. serial numbers, of all stolen or lost mobile phones that must be prevented from using the system. All users are listed as either Black, i.e. non-valid users, White as genuine subscribers and Grey when under suspicion or observation.

Um: This is the air interface between the MS and BTS.

Abis: This is the interface between the BTS and the BSC.

5.1.2 The GSM Subsystems

The network components are arranged into three subsystems, i.e. the mobile station, the base station subsystem and the network sub-system (see Figure 5.3).

The MS consists of the mobile unit or terminal with a SIM smart card. The SIM card (see Section 5.2.2) has access protection in the form of a password or PIN of usually six digits with an error counter of three, which, when exceeded, allows no further entry attempts and blocks the card, thereby rendering log on to the system impossible. The card carries quite a variety of information apart from the 'phone card' memory. This includes the IMSI, which is used to identify the user to the system. The card itself is 'mobile' in that the owner is able to insert it into any compatible hand phone hardware and utilise the system. The SIM card, more

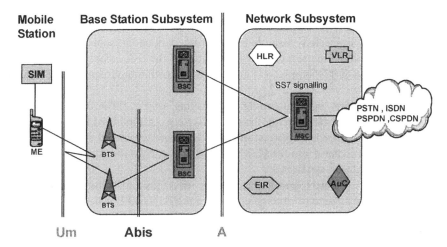

Figure 5.3   The GSM subsystems

specifically the IMSI, is the remote partner in the GSM security protocol as it contains the secret key for the user authentication. The MS also carries the fixed A5 algorithm that is used for the call encryption over the Um interface.

The base station subsystem consists of two major components:

- The BTS and the BSC communicate with each other over the *Abis* interface. The BTS carries up to 16 RF transceivers for its cell and therefore handles the signalling between the MS and itself over the 'Air' or Um Interface.
- The BSC, as it name implies is the controlling element of the subsystem, and it is responsible for one or several BTSs. It handles the RF channel set-up, the spread spectrum frequency hopping, and call handovers between cells, and routes mobile calls to the MSC when required. A call made between two mobiles located within the same cell can be controlled by the BTS and BSC.

The network subsystem consists of four major components:

- MSC
- VLR
- EIR
- AuC
- The MSC, which is the mobile equivalent to the normal PSTN switching centre and routes calls to and from the mobile subscribers and the normal telephone network. It controls if the IMSI and SIM card are thought to be the remote partner in the GSM security protocol; then, the MSC with the location registers, EIR and the AuC can be considered to be the central partner in the proceedings. The location registers will be discussed in greater detail in Section 5.2.1 but for the moment can be simply considered as facilitating the call routing and roaming operations. As mentioned above, the EIR is a database that contains a list of all of the mobile equipment using the network. Each mobile is identified by its own IMEI, and this enables the network to be able to monitor its users and to allow valid users to use its facilities whilst preventing stolen or non-approved units from being used. The AuC, the Authentication Centre is a protected database that stores a copy of the encryption key used in the authentication process and encryption over the Um interface.

5.1.3 The GSM Radio Um Interface

The original GSM RF band is in the 900-MHz band with the uplink, i.e. from mobile to BTS operating between 890 and 915 MHz and the downlink BTS to MS, using the 935—960-MHz band. With a single channel having a 200-kHz bandwidth, the frequency division multiple access (FDMA) gives 124 channels for calls over the 25-MHz band. This is further expanded by time division multiple access (TDMA) to give a massive 992 duplex channels of telephone traffic.

Other important features of the radio channel include:

- Adaptive time alignment, which enables the MS to correct its transmit time slot to compensate for the propagation delay.
- GMSK modulation, which provides both spectral efficiency and low 'out of band' interference.

- Discontinuous transmission and reception, which is the turning off of the MS during idle periods within transmission and receiving conditions. By this technique, the MS battery performance is improved, and also co-channel interference is reduced.
- Slow frequency hopping, is a spread spectrum technique (see Chapter 7) that helps to counter fading and co-channel interference. It is quite probable that in some locations, e.g. built-up areas, radio channel fading will occur within the operating frequency band and rather than specific cell channels suffering as permanent noisy channels, all the channels in use will go through a frequency hopping routine so that for a short period of time, each will be subject to the vagaries of the poor channel. Hence, the spread spectrum technique spreads the interference over the whole range of channels, rather than seriously degrading just a few of them. Note that the slow frequency hopping is not intended primarily as a security device.

5.2 Standard GSM Security Features

As implied earlier in this chapter, the standard security features of the GSM system (Figure 5.4) include:

- The AuC
- HLR
- VLR
- SIM cards
- IMSI & TMSI
- Encryption
- TDMA
- Frequency hopping
- EIR/IMIE

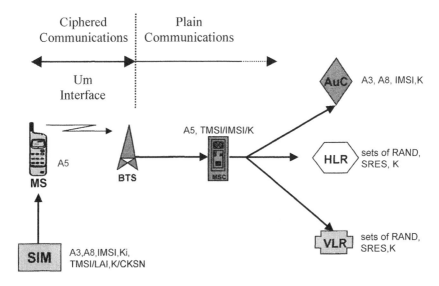

Figure 5.4 The distribution of GSM security elements

5.2.1 The AuC

From Chapter 1, the reader is able to appreciate that the *authentication* of messages and users is an essential aspect of communications security. This is no truer than in GSM, where, as in all radio operations, the transmission medium is freely available for all interested parties to use and monitor. The authentication centre and its supporting unit, the HLR, seek to remove the threat to authentication. The AuC provides the HLR with several parameters in order to enable the authentication of a mobile user.

The AuC carries all of the algorithms required by the networks and knows which algorithm to use in authenticating a specific customer. Therefore, the AuC must be protected from abuse and attack. The SIM card also carries the user's algorithms and so is also in need of access protection. The process is illustrated in Figure 5.5.

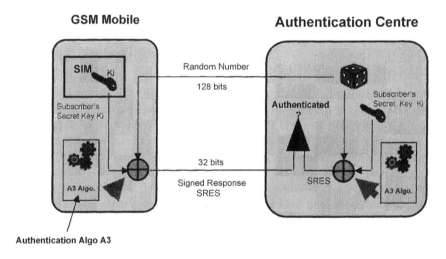

Figure 5.5   The GSM authentication process

The AuC checks the validity of the user's SIM card using the A3 algorithms stored on the SIM and the AuC. It does so by using two inputs, an authentication key (KI) and a random number (RND) of 128 bits, that are transmitted by the network to the mobile over the Um interface. This RND is received by the MS being validated and is passed to its SIM card. The SIM card authentication key is then used by the A3 algorithm to encipher the RND and subsequently generates an output, the 32-bit SRES. The SRES is retransmitted back to the AuC where it is checked against the expected result calculated by the AuC. An agreement here thus authenticates that MS as a bona fide user. Any invalid user would not be in possession of either the correct KI or the A3 algorithm and therefore would not be able to calculate the correct SRES response. The random number generation ensures that the SRES is different on each login occasion. This is a typical example of challenge/response system.

5.2.2 The HLR

Each GSM network has one home location register. It holds a number of sensitive parameters, including the mobile user's details such as billing, the A3 algorithm for the authentication, the A8 algorithm for message encryption and the corresponding encrypting key Ki. It is also responsible for the generation of the random number sequence used for the authentication.

Being the host of so many critical parameters, the HLR is the target for many an intruder, and if it is not the usual cryptographic security features that are of interest, then it might well be fraud and the billing procedures where he seeks to benefit.

5.2.3 The VLR

The visitor's location register contains the details of all the mobile units located in the corresponding GMSC. Whilst the HLR carries all of the permanent subscriber details, the VLR contains the TMSI of the relevant MS, and this is used for Um signalling rather than the IMSI, which must be secure from eavesdroppers. The VLR also indicates to the system the actual physical location of the mobile and supports the authentication procedures with the GMSC, when a visiting MS first logs on to a different network.

5.2.4 SIM Card

The SIM card or subscriber identity module is a smart card with a processor and memory chip. The SIM card is at the heart of GSM security, it is crucial to authentication procedure and the signal encryption process. The card contains the IMSI, both of the algorithms for these functions, i.e. the A3 and A8 algorithms, respectively. In addition, the SIM carries the Ki, the user's individual authentication key and of course its access control element, the PIN.

SIM access control relies upon the possession of the SIM card itself and knowledge of the personal identifying number. A user forgetting their PIN or a 'brute-force attack' from a dishonest party is confronted with an error counter allowing three erroneous attempts to access the card before it becomes blocked. A blocked SIM card can only be unblocked by obtaining and entering the eight-digit personal unblocking key (PUK). This is usually available from the service provider, who is able to verify the genuine user after a verbal interview.

Apart from the security parameters, the SIM card contains the subscriber's personal calling details such as:

- Personal phone numbers
- Subscriber ID for the network, i.e. the IMSI
- Short message memory
- Details to facilitate roaming during international travel
- Billing details

The information flow between the SIM card and the networks VLR is shown in Figure 5.11.

As the mobile phone industry moves into the third generation to implement WAP and GPRS systems, SIM cards, as they now are, will be insufficient for the more complex operations that will be required of them. Whilst on-board phone memory capacity will undoubtedly improve, the SIM card will still be required to carry the major security features especially for payment of services as e-commerce becomes more popular. It seems probable that two-slot hand phones will be developed to enable the highly desirable security strategy of having easily removable SIM/smart card facilities. The development of 'permanently on'

GPRS networks will basically require a permanently installed SIM card, which invites security problems. It would make much more sense to have a second removable card that will only be inserted into the hand phone when an e-commercial transaction is being carried out.

At the present time, the security of the SIM card relies upon symmetrical encryption, but in the future, as mobile phone technology advances and SIM cards develop to match these advances, more emphasis will be made on the card as a tool of authentication. The most obvious answer to this requirement is to use public key systems where both the private key and the public key parts are required for any transaction to occur. The smart card is an ideal tool for this application.

5.2.5 The IMSI & TMSI

The IMSI is the subscriber's international mobile subscriber's identification.

The TMSI is the temporary mobile subscriber identity. This is issued to a roaming visiting user after its authentication and encryption procedures have taken place, by the local VLR. The mobile responds by confirming the reception. The whole procedure is secured by the use of the A5 encryption algorithm (see Figure 5.6).

The TMSI is used to identify that subscriber during their stay in the region of that particular VLR, and its issue to the subscriber helps to maintain the user's IMSI confidentiality by protecting the IMSI from any eavesdropper on the radio path. It is retained for the period of that stay, and it is changed from time to time during the handover protocol. It is stored on the visitor's SIM card for possible use again during subsequent log-ins within the same region. For calls outside the location area, the local area identification (LAI) is necessary in addition to the TMSI. Hence, subscribers are permitted to make calls and update their locations without the necessity to reveal the all-important IMSI and therefore prevent a subscriber's location being available to any listener of the signalling over the Um interface.

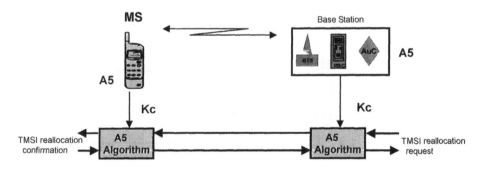

Figure 5.6   The TMSI application

5.2.6 Standard GSM Encryption

There is little doubt that the GSM system is far more secure than its predecessors, competitors and the fixed subscriber PSTN system. The digitised speech from the vocoders, the Gaussian minimum shift keying (GMSK) modulation, frequency hopping and the TDMA multiplexing all add up to deter all but the most persistent of eavesdroppers. However, the main problem with GSM as far as security is concerned, as hinted at by Figures 5.10 and 5.11 showing the extent of encryption through the signal routing, is that the only parts of a GSM transmission that are encrypted are the radio channels between the MS and the BTS, i.e. the Um interface. The remainder of the signal path to either a fixed telephone subscriber or to another GSM located in a different cell is over the public telephone system, where normally no protection confidentiality exists. Therefore, any eavesdropper need not worry about attacking the semi-protective measures inherent in the GSM system because everything is in plain within the network side of the BTS elements. Naturally, any attacking body will exploit the weakest link, and the PSTN or ISDN organisms are just that.

Consider Figures 5.7–5.12, which show the voice ciphering process.

Immediately after the SRES signal and subscriber authentication, the VLR commands the MSC to control the BSC and, subsequently, the BTS, in cipher mode. The Kc data, which have been derived from the Ki and A8 algorithm in the HLR, are despatched to the BTS via the BSC, at which point, the BTS commands the mobile (MS) to switch to cipher mode. The MS, or more specifically the SIM card, which contains the subscriber's key Ki, applies this key data and the RAND used in the authentication process to the A8 algorithm. The result is the output of the Kc, the 64-bit ciphering key, which is then fed to the mobile's A5 algorithm that generates the keystream to cipher the digitised voice signal in transmit mode and deci-pher the received voice signal in receive mode. During this period, the BTS, after SRES authentication, has also switched to cipher mode and similarly uses the Kc key to encrypt the voice signals of the same channel. Hence, the call over the Um interface is encrypted and therefore confidential between the MS and the BTS.

From a cryptographic point of view, the algorithms A3 for user authentication, A5 for message encryption and A8, the latter being a key-generating algorithm supporting the A5, are quite weak compared with other standards. Both the A3 and A8 algorithms are imple-mented in the SIM card along with the subscriber's unique key (Ki) and are therefore transportable in a secure manner when the user is roaming within other operator's networks. The A5 algorithm is a fixture in the GSM mobile hardware and is a stream cipher using three linear feedback shift registers (see Figure 5.8) to give an effective key length of 64 bits. The

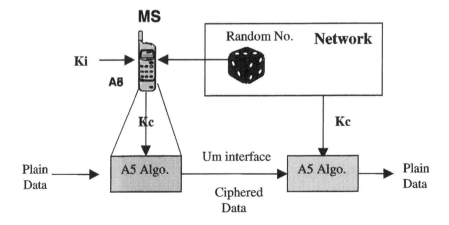

Figure 5.7   The basic encryption process

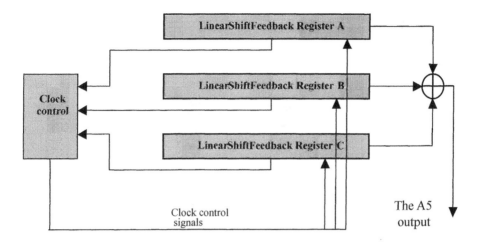

Figure 5.8   The A5 keystream generation

KC session key of 64-bit length is loaded into the registers, which are clocked through for a short period, and then 228 bits are taken as from the output to form the keystreams for the uplink (114 bits) and downlink (114 bits) encryption.

There are countless claims and rumours of possible 'breaks' into GSM algorithms by various bodies, but at the time of writing, none was substantiated in the author's opinion.

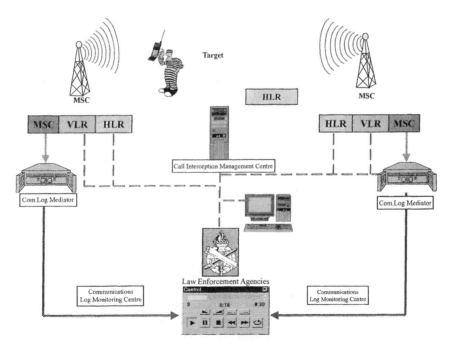

Figure 5.9   A GSM monitoring centre

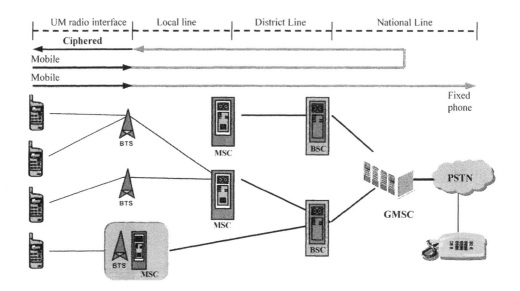

Figure 5.10   The extent of standard GSM encryption

However, the weight of attacks brought to bear on the GSM security, some by irreproachable institutes, leads one to suspect that there is much truth in the claims made. It probably never ceases to amaze the informed, casual observer the degree of attention that GSM encryption has attracted and how much of it is in the public domain. Strictly speaking, though, most manufacturers of cryptographic equipment actually assume a worst-state scenario in which the algorithm is actually in possession of the opposition anyway. This assumption is also known as 'Kerckhoff's Assumption', and all modern algorithms and security equipment are built around it. By what means is unimportant. Instead, the security is based upon the heavy reliance on the strength of secret keys and frequent changes of them. Sadly, at the present time, the Ki in GSM is a semi-permanent key and therefore somewhat vulnerable. As threats grow in sophistication and number, more regular Ki changes would certainly be beneficial to the systems security but, at the same time, would create a headache for those responsible for the key distribution. Having said that, when one considers the Ki as being a 128-bit key, the examples of 'Exhaustive Key Search' suggested in Table 2.2 support its strength, but the statistics point the finger at the Kc key, which is only a 64-bit key.

A further point of interest is that all of the GSM algorithms were developed in Great Britain for ETSI and the original GSM founder group, and are therefore subject to the thorny question of export restrictions. In view of the possible development of the GSM system to other regions, three levels of channel security involving the A5 encryption algorithm were developed, designated A5/1, A5/2 and 'unencrypted'. The highest security version A5/1 was for use within the founder group and in countries, which are members of Conference of European Posts and Telegraphs (CEPT). The second level, A5/2, is for export to nations falling outside that category, i.e. most countries including those of central and eastern Europe. The final level has no encryption at all for specially designated countries such as Russia.

It is also uncertain whether a particular service provider, even if they have the right to use

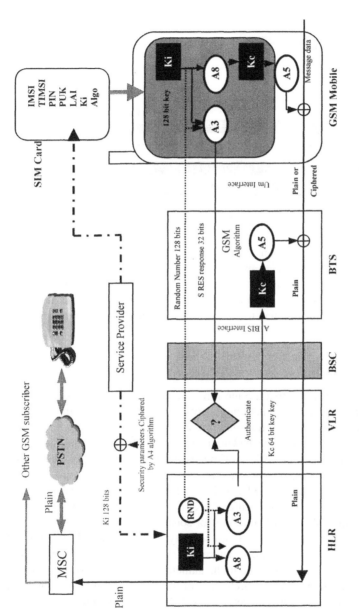

Figure 5.11 The complete authentication and encryption process with processing elements

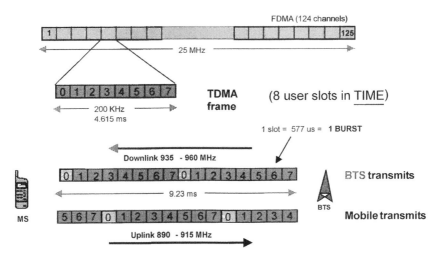

Figure 5.12   TDMA structure

encryption, will in fact implement it at all. In some countries, it is a simple task to find out simply by calling the local service and asking. Some are happy to broadcast the fact that they do use encryption, but others, at least in the author's experience, will refuse to comment at all on the subject. Many GSM providers are forced to follow government edict and have little choice in the matter, but one has to be circumspect when no confirmation is forthcoming following a polite enquiry.

5.2.7 Cryptographic Attacks on the GSM Algorithms

From Chapter 2, we can see that a brute force attack on the A5 and A8 algorithm, with the given constraints, would take something just over 20 h of computer time, which makes real-time analysis of the Um interface transmissions an impossibility. It may, however, be possible to record transmission frames over the Um interface for subsequent analysis at a later time. In encrypting the voice data, the A5 algorithm is initialised by the Kc, the session key, but produces a different keystream for every frame. As the frame numbers are readily available to the eavesdropper, knowledge of the Kc will enable the attacker to decipher the voice message frame. To limit the vulnerability, a different Kc is generated on the event of a new authentication, but the same value could be in use for some time.

The most important component in GSM security is the Ki, the subscriber's 128 bit secret key. It is generated at the subscription centre and stored on the subscriber's SIM card and is used as a symmetrical encryption step in the authentication and A5 voice encryption. The Ki is transmitted to the HLR under the protection of the A4 algorithm. Deriving the Ki from the A4 algorithm would enable the attacker to clone SIM cards and monitor any call made by the unfortunate subscriber possessing that particular Ki. There are protection procedures in the GSM system to detect the simultaneous use of two Kis of the same value, and in those circumstances, the account is usually terminated immediately. However, the use of a Ki simply to monitor calls is undetectable by the system, and its knowledge in the hands of any organisation is the key to monitoring any calls made by, or to, the subscriber in question.

At no insignificant cost, GSM monitoring equipment is available to those authorities such as government, law-enforcement agencies and others wishing to eavesdrop on GSM subscribers. The principle of just such an eavesdropping centre is illustrated in Figure 5.10. The agency will target the individual by entering their mobile telephone number, amongst other details, into the management centre. It can be arranged that the management centre has access to the local operator's HLR and VLR, so that when the subscriber either makes a call or is called, the centre logs that call. Monitoring takes place in the form of a conference call, with the monitoring station purely playing a passive role. A typical centre can monitor many simultaneous calls throughout 24 h a day, 7 days a week, with ease.

5.2.8 TDMA Time Division Multiple Access

For the GSM system to cope with the ever-growing number of subscribers, two levels of multiplexing are used. FDMA has been outlined in Section 5.1, for the sake of the inquisitive or uninformed, and for completion, TDMA is briefly mentioned here (see Figure 5.12).

The FDMA divides the 25-MHz band into 124 carrier frequencies spaced at 200 kHz. One or more carriers are allocated to each base station. Each carrier frequency is then split into time segments by TDMA, known as bursts. The length of a burst is 0.577 ms as each TDMA frame is made up of eight bursts, so the frame period is 4.615 ms. Each burst carries one physical mobile channel. The product of TDMA with FDMA gives a total of 992 duplex GSM channels. The TDMA frame is itself a component of a hierarchy of frames including Multiframe, Superframe and Hyperframe.

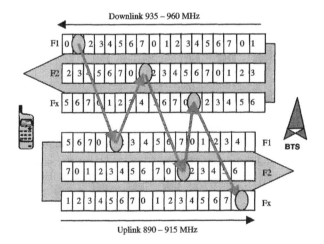

Figure 5.13   Slow frequency hopping in GSM systems

5.2.9 Frequency Hopping

Slow frequency hopping is used in GSM systems, not primarily as a security element, but rather as a spread spectrum technique to overcome the effects of radio channel fading, which is typical of built-up locations. However, any party wishing to monitor the transmissions of a hopping channel would need to be in possession of the hopping algorithm. From Figures 5.13

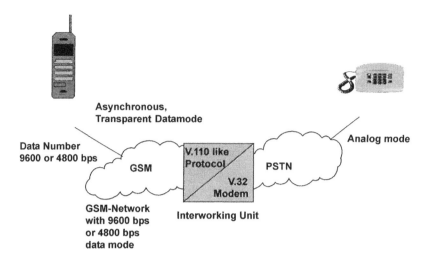

Figure 5.14   The requirements for secure GSM operation

and 5.14, it can be seen that the BTS is transmitting to the MS on channel 1 at frequency 1. The MS also transmits to the BTS on the same channel and frequency, but at a different time. Together, the MS and BTS hop as a single physical channel over the available bandwidth, which, for simplicity's sake, only covers three frequencies in the diagram. The result is that each channel is not seriously affected by any fading of a fixed frequency within the spectrum. It should be noted that the hopping algorithm is not controlled by any cryptographic key but is controlled by the BSC, which agrees a particular hopping sequence with the MS.

5.3 Custom Security for GSM Users

It can be seen from the discussion above that the only possibility for a user to have complete confidentiality over a GSM system is to employ a point-to-point encryption strategy. With such a strategy, the subscriber would enjoy assured privacy whether the calls are simply between GSM mobiles or GSM to PSTN stations. The techniques involved in achieving such a goal are not so simple to implement, and throughout the world of telecommunications, only a handful of cryptographic manufacturers produce the necessary hardware. As this is very much a niche market, the cost of outright mobile telephone security takes it well beyond the pocket of the man in the street and even most businesses. Therefore, the purchase of such equipment is limited to government agencies, armed forces, royal families and senior staff members of world-wide bodies such as the UN.

Having decided that custom security is the way to go, the first problem is that to use it on the GSM system, one must have an operator who is willing to allocate a GSM data channel for each of the mobile users within the secure group. A 'data channel' can be thought of as a second line to the mobile phone, a line with a different subscriber number to that of the voice channel, which will be configured to carry data as per computer data, rather than the normal voice channel. Unlike the voice channel, the data line has no error protection protocols and is therefore often referred to as a transparent channel. This is a mandatory requirement for custom security.

In addition to the requirement for a transparent data channel, most operators, especially those within Western Europe, do offer the data channel to their subscribers. However, some providers do not. The reasons for this are usually that there is insufficient demand, the operator does not have the infrastructure to support the extra channel or, possibly, that they wish to prevent subscribers utilising the facility for such operations as encrypted voice calls.

Obtaining the use of a transparent channel is only the tip of the iceberg as other, not so obvious, protocol layers need to be considered. It appears that the sole specification of the bearer service number BS26 for 9600 bps and BS25 for 4800 bps, as defined in the GSM technical specifications 02.01 and 02.02, is in some cases inadequate to ensure a complete set-up of the network bearer service. Beneath every bearer service number is a set of lower layer attributes called 'bearer capability elements', which are defined in the GSM technical specification 07.01. These attributes specify a certain bearer service number in detail, and it is possible that the default values may differ between different infrastructure suppliers. If these default values do not reflect the proper settings as required by a customised secure GSM mobile telephone, ciphered calls might not be possible, even though the bearer services BS25 and BS26 have been assigned to the users requesting them. Detailed verification with the service provider's technical set-up must ensure that all of the attributes, i.e. sub-elements are set up as the secure GSM manufacturer specifies.

One ploy, which might be a solution to provider reluctance, would be to carry a spare SIM card provided by a co-operating GSM provider, possibly a foreign provider, and exchange it with the existing SIM. This means, of course, having another telephone number to deal with, but it would be a small cost to pay for high security.

The GSM organisation publishes a web page by the name of 'GSM World' on which 'almost all' the facilities, including information on roaming agreements of all members, are there for all to see. However, not all member operators are totally honest about what they publish on the page as experience has shown that whilst some do not advertise it, but have it, others advertise it but cannot provide it. The reasons behind this fact are left to the imagination of the reader, as it is a very sensitive issue. One would expect, though, that government-sponsored agencies wishing to use the channel would be able to overcome any access problems.

Once assured of access to a GSM 'data channel', the would-be client might expect to require the following specification, ignoring the usual cosmetic mobile facilities:

- End-to-end voice encryption
- Strong encryption algorithm with long keys, e.g. 128 bits
- Tamper-resistant security module
- Encrypted key storage
- Multi-key facilities that would enable the design of a secure network structure
- Up-to-date key management techniques and tools
- Emergency clearing of key and sensitive data
- Access control
- Normal GSM 'plain' operating mode as an option
- Excellent voice quality with speaker recognition
- Easy to use

5.3.1 The Custom Encryption Process

The basic structure of an 'off the shelf' GSM hand phone is worthy of some discussion (see Figure 5.15), and its major elements are as follows:

- The *antenna combiner* couples the receiver and transmitter into a single antenna.
- The *receiver* contains the 'front end' signal processing, incorporating a filter and a mixer to produce the intermediate frequency (IF). The signal strength of this is measured and displayed to the user.
- The *equalizer* compensates for the distortion due to multi path fading.
- The *demodulator* extracts the bit stream from the IF stage.
- The *demultiplexer* uses the frame numbering to sort out the received information from the appropriate time slots into logical channels.
- The *channel codec* codes or decodes a bit sequence to and from the multiplexer, i.e. it processes the time slots containing the speech data and also the control channels and frame package overheads. It passes signal frames to the control signalling unit and speech frames to the speech codec.
- The *speech codec* either rebuilds the human speech characteristics from the 260-bit speech blocks, passing the digitised speech to the digital-to-analogue converter (DAC) or, in transmit mode, compresses the digitised speech signal from the analogue-to-digital converter (ADC) into the 260-bit blocks before encoding them.
- The *control and signalling unit* performs all control functions such as power control, channel selection and various other functions.
- The *burst building unit* builds the channel burst structure and multiplexes the channels into the frame structure. The digital signal is then converted into an analogue signal for the transmitter stage.
- The *transmitter consists* of the IF and RF stages including filters and amplifiers to control the output power according to the demands from the BTS.
- The *VCO and synthesizer* provides all mixer frequencies and clock signals.

The essential difference (see Figure 5.16) between the original GSM mobile and the customized version is that the standard voice codec, typically an regular pulse excited–linear

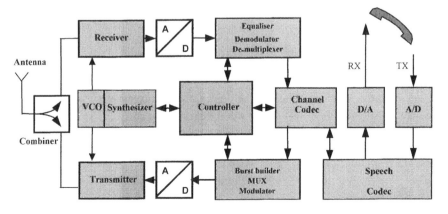

Figure 5.15 Block diagram showing the modules of a basic GSM mobile unit

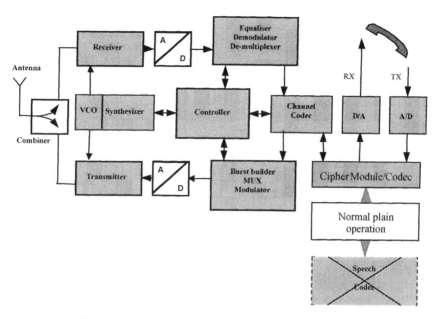

Figure 5.16   Block diagram of a customized secure GSM

predictive coder (RPE-LPC) device, is bypassed to eliminate the error correction processes normally found in the standard GSM, hence the term 'transparency'. The original codec is replaced with a cipher module that incorporates a different coder such as an AMBE Vocoder chip. It is the codecs that are largely responsible for the high-quality voice recognition found in the GSM telephones. The AMBE vocoder chip is a digital signal processor (DSP)-based full duplex voice coding solution for voice compression applications with superior functions to the linear predictive vocoders. The digitised voice signal from an external ADC is converted into a compressed data stream that is encoded by the vocoder chip and then fed to the output interface for transmission. Simultaneously, the chip receives compressed digital data from the remote side of the channel and then reconstructs the original digital speech signal before passing it out to a DAC. The coder incorporates FEC error protection and gives good-quality speech with high bit error (BER) rates of up to 5%, which makes the device ideal for mobile communications.

The typical arrangement for the ciphering process is shown in greater detail in Figure 5.17. In transmit mode, the output of the digital audio interface, which, in a 64 kb/s format, is passed to the speech codec otherwise called a DSP, i.e. the AMBE coder, which processes the signal to a compressed 5.6 kB/s and then mixed with the ciphering keystream by the modulo 2 addition in an exclusive OR gate. The ciphered TX signal is repackaged to include synchronizing information before being passed to the normal GSM transmit operation, through an interface, typically an RS-232. The receive mode is the mirror image of the transmit ciphering.

The set-up of a ciphered call between two custom units (mobile or compatible desktop phones) would follow the following sequence of events.

The first step would be to set-up a plain call in the normal way but by dialling the number of the transparent data channel belonging to the recipient. Once a voice communications

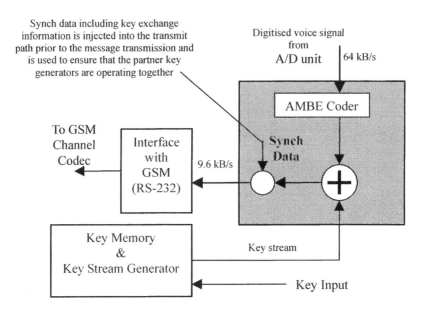

Figure 5.17 The ciphering process

link has been established, the calling parties will elect to move from plain mode to cipher mode.

On the command to initiate the change over, the two stations will set about negotiating a suitable key for that link. The participating units will search their key memories for a common key with the same key value and other compatible parameters. The method of determining an appropriate channel key will vary according to which method of encryption is used, but it will include a check of the key signatures, the key label or ID and the validity of the key. The validity of a key will probably involve the period of time for which a certain key is active and the secret key selection process.

Depending upon the key management strategy adopted, the key structure may be hierarchical with priorities based upon the network configuration, or it might be link-dependent. As the encryption is most likely to be symmetrical, due to the fact that asymmetrical is more difficult to implement and is slower, the same key at each end, with matching data values, will be selected.

The actual switch over time from plain to cipher mode will depend on the time taken for the key negotiation and the synchronisation to be completed, at which point, an alarm or indicating LED will notify the users to continue their conversation in cipher mode. The synch process may well take a few seconds to complete, and, whilst the key agreement accounts for some of this time, by far the greatest delay is because of the modem training.

The switch to cipher mode will take place on the instruction of one of the calling parties, but for the sake of security, a switch back to plain mode must only be on the command of both parties, otherwise it might be possible for one side to secretively change the mode of operation, with its consequences, without the remote party being aware of the change. To further prevent this possibility, a custom secure system should at least provide the option of including an audio 'plain marker' alarm and flashing LEDs that give a clear and continuous indication of the operating status.

5.3.2 Key Systems

From Chapter 2, it can be seen that there are several alternatives when it comes to selecting a particular key system. Very often, the nature of the application governs the preferred choice, and with voice encryption systems, symmetric encryption is preferred. Within this technology, there are further alternatives:

- Symmetrical session key (direct) encryption
- Symmetrical session key (derived) encryption
- Asymmetrical session key encryption

This chapter will explore the session key variants that may be offered by various manufacturers. Previously, it was stated that session keys do impose a greater security than fixed keys as the former are changed frequently according to each communication, and Figure 5.20 illustrates a prime example of the application.

5.3.2.1 Symmetric Direct Key Systems

Consider Figure 5.18, where the initialisation vector (IV) is a random sequence, typically of hex characters that would be truncated down to form a 128-bit binary value, the purpose of which is to ensure that the starting position of the key generators of the encrypting algorithm is at a different point for each operation, thereby removing repetitive cycles from the process. In this instance, each unit is contributing part of the IV to determine the key generator initial set-up.

The secret key is made up of 32 characters, and as the example here is one of a symmetrical

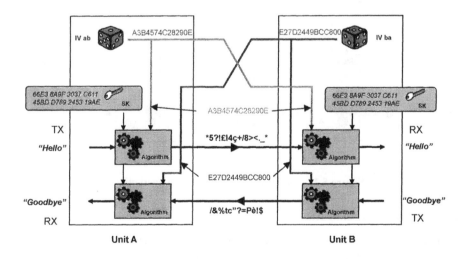

Figure 5.18   Direct symmetrical encryption

cipher, the secret key (SK) must be the same in both units. This key is then fed to both the RX and TX key generators to cipher the duplex digital voice signals.

This method of encryption requires the least amount of computing power as the confidentiality of the channel is ensured by the secrecy of the secret key and the algorithm. However, it does not guard against the threat of 'replay' attacks in one direction, e.g. Unit A to Unit B, but in the other direction, Unit B will generate a new random IV and so avoid the threat. It is best for small networks where regular changes of key data are more easily implemented, and it can be used for securing data to be archived.

5.3.2.2 A Derived Symmetric Session Key

This method, as described in Figure 5.19, is more complex than the previous option in that it develops a session key through an interactive key agreement protocol and uses a random initialisation vector. The confidentiality of the method depends upon the secrecy of the SK and the algorithm with the added advantage that the SK is not directly used to encrypt the voice channel.

Replay attacks are countered by the fact that new keys are generated by both parties for every session. The transmission of a recorded message will therefore not be possible, as it will have been encrypted by a different session key.

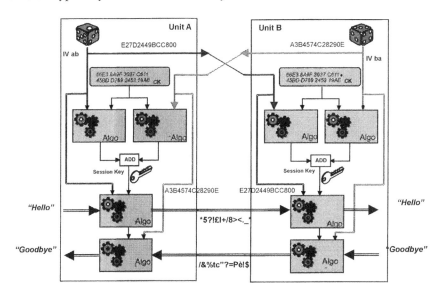

Figure 5.19 Derived symmetric session encryption

5.3.2.3 Asymmetric Session Key

As an alternative, asymmetrical ciphers are used to generate a cipher key composed of two different values (see Chapter 2). Using the Diffie–Hellman key exchange technique, both mobile units generate their random public session keys, which are passed, in plain, to the other mobile. In each case, the received value is then used with that unit's own public key value to calculate the cipher key. Authentication is provided by the inclusion of a symmetric key.

The message confidentiality is based upon both the asymmetrical derived key and the stored symmetric key. As with the symmetrical derived method, the secret key (SK) is not used directly to cipher the plain voice data stream and therefore makes it less vulnerable to cryptographic analysis.

5.3.3 Cryptographic Parameters and Algorithms

A high-quality security system would use multiple algorithms. The first for the actual message data encryption, another for the tamperproofing and a third for the key management, whether it be off line or on line. As discussed in Chapter 2, customisation of algorithms is very difficult to come by but is certainly a desirable facility if it can be bought.

A tamperproofing algorithm and associated key are used to protect sensitive data normally resident in the security module. Such a device would probably be a permanent randomly generated key, of which no copy should exist, and this makes it impossible for the attacker to gain access to the tamperproof key or the data it would protect. Most data can be protected by encrypting it under the 'tamperproof key', but the tamperproof key itself cannot be so protected. Therefore, this key must be protected physically.

Typically, a system buyer would be looking for a 32-character key structure, giving 128 bit keys. The larger the proposed network, the greater would be the number of keys required to securely configure the topology, so the provision of around 100 SK storage capacity is practical. Whilst 'default keys' can be useful, e.g. when other keys have been lost, there is a tendency to rely upon them too much, and it should be remembered that they are only intended to give secure communications, albeit of reduced quality, in cases of emergency. A manufacturer's default key is most likely the same for all customers, so the danger is obvious, and the default key should only be used as an emergency and with care. It should only be enabled at the discretion of the manager and should be clearly indicated by displays and alarms when in use.

The key stream period should be quoted and should be very large, typically in excess of 10^{20} years (see Section 2.2).

5.3.4 Security Architecture

For the very best security, once the secret keys have been loaded into the unit, they should never leave the security module in plain. If keys or any other sensitive parameters have to be moved around the unit, they should be ciphered by a separate key, i.e. the 'storage key'. Wherever possible, all security parameters should be enclosed in a tamperproof module that prevents the readout of secret keys, hence protecting them against coping and modification. The reading out of the algorithm is not a problem in equipment, which employs standardized algorithms such as 3DES or AES. The protection of an algorithm is only really an issue in high-security, military or governmental ciphering equipment that uses secret algorithms.

The security module itself may either be constructed as an integral unit of the GSM hand phone or as a separate detachable module that can be connected to the mobile's interface

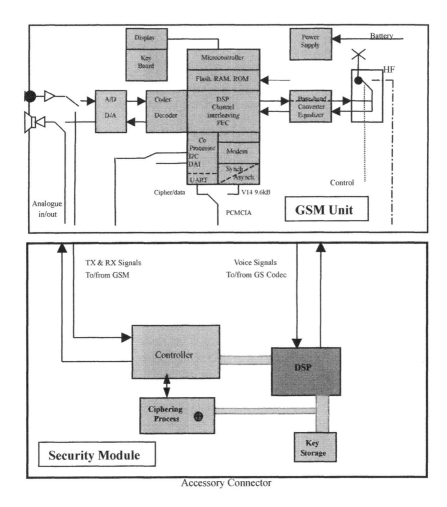

Figure 5.20   Typical security module with standard GSM architecture

located at the bottom of the unit. This option could be very useful, as the encryption module can be removed and stored in a secure location whenever it is not required.

Wherever the cipher module is located, there should be strict Red/Black separation of components (see Figure 5.20) so that there is no trace of plain data of any kind detectable within the cipher circuitry and likewise no sensitive parameters from the cipher unit detectable within the plain circuits.

5.3.5 Cipher Unit Hardware Elements

- Security processor/controller

    ROM – for secure algorithm with no readout facility
    RAM – working memory

- Flash EPROM containing:

  Program for DSP operation
  IT MIB for operating parameters
  Security MIB – for keys and client keys, etc.

- Bi-colour LEDs

  Red – alarm or error indication
  Green – ciphered communication in process
  Off – plain communication in process

- DSP

  DSP operation
  Exclusive OR ciphering function

5.3.6 System Overview with Secure GSM and Fixed Subscriber Equipment

Figure 5.21 not only illustrates a custom GSM/GSM link or network but also introduces the probable need for the custom GSM mobile to be compatible with desktop stations. After all, it is point-to-point security that is required, and it is almost inconceivable that a secure network will consist solely of mobile phones. Therefore, the desktop units must have the same security system as the GSM telephones.

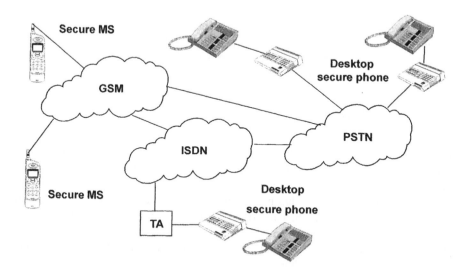

Figure 5.21   Secure system overview

5.4 Key Management and Tools

Password access should have at least two levels, one for the operator with limited access to menu operations and a second for the security manager giving them full control over the operating and encryption parameters. As in most cases, failure to protect the password hierarchy might have catastrophic consequences for the network's security, and hence the manager's access must be held as top secret.

5.4.1 Key Distribution and Loading

The limited size of hand-held telephones makes them largely unsuitable to adapt for key input operations, and although security data input through the manual operation of the key board is possible, the human factor often introduces errors that are difficult to combat at a distance. Wherever possible, and especially if the operators are neither so technically minded nor security aware, it makes sense to avoid the user inputs of such sensitive data wherever possible. Just consider the frequency of 'wrong number' dialling when contemplating a team of secure GSM carriers entering their own keys and security parameters. This method of key entry still relies upon the key distribution and for human input, and at some stage, the key data have to be available in plain text and so present a highly undesirable state of affairs.

At least one manufacturer takes advantage of not only producing fixed phones that are cryptographically compatible with the mobiles but also making major use of them as distribution centres in the key and network management philosophy. With more space available for key distribution hardware and software, the desktop telephones are able to accept the down line transmission of secret keys from a management centre. These downloads, of course, must be protected by transfer, or transport encryption. The security parameters and key data can then be transferred to the GSM telephone either through a cloning cable connected between the desk top unit and the mobile's accessory interface, as identified in Figure 5.20, or by a chip card transfer between the two devices.

5.4.2 Chip Cards and Readers

Another alternative is to use chip card distribution of security parameters and key data directly from the management centre. This is much more secure and reliable than pure manual input. All that is required of the user is to connect a card reader to the GSM interface and input a password to gain access to the load function as directed by a basic menu on the phone's display, and if the function has been enabled by the security manager, the loading of keys and security parameters can be carried out in an anonymous fashion. The use of chip cards and down line loading through an intermediate desk top secure telephone station, or a combination of both methods, provides a number of alternatives that can be tailored to most situations. As ever, a package of alternatives infers greater efficiency and flexibility than a single solution to key distribution.

5.4.3 Key Signatures

As discussed in other chapters, key signatures form an essential part of key management and

distribution. There are a number of ways to construct key signatures from relevant parameters, and one alternative suggestion is offered here.

Whereas, in other applications, only the actual key data were used to construct the signature along with the algorithm, in this alternative, the key data are supplemented by information about the link, e.g. Lon–Wash and the validity period of the key.

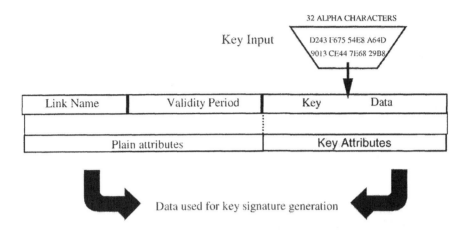

Figure 5.22   Key signature components

5.5 GPRS General Packet Radio Systems

One cannot talk about the security of the GSM system without peering over the horizon and
visualising the problems that will arise with the advent of GPRS. GPRS represents a major
step in mobile communications, from the relatively simple GSM voice application into the
complex world of data and web access. It is a step that will see the mobile terminal being used
as a 'phone-shopping' tool, a device for making e-payments. A device that can be used as an
electric wallet, which requires the periodic topping up of cash deposits by transmitting the
user's PIN over the GSM system. So, it is not difficult to imagine the pitfalls that await both
the user and provider of GPRS, especially when one considers the leaky security of the GSM
organism. However, it is not all dark clouds ahead as the nature of GPRS transmission does
import some inherent security features into the GSM operation. After all, the general consensus is that despite all of its cryptographic frailties, data over the Um interface are considered
to be confidential for real-time operations.

5.5.1 Basic GPRS Operation and Security

GPRS is a system that uses the GSM structure, with modifications, to implement high-speed
data transmission between the mobile station and the other part to the call. It achieves the high
speeds by transmitting packets of data in a 'parallel transmission' of data slots that utilize not
a single BTS, as with a GSM call, but a number of BTSs simultaneously (see Figure 5.23).
Each packet is transmitted to a BTS, which sees each individual packet as a different
call and so does not care about the order of the packets, i.e. the message from the mobile

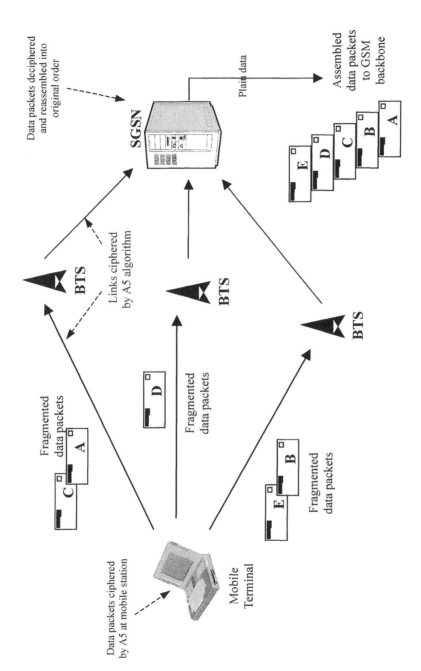

Figure 5.23   Basic GPRS operation

station is defragmented by the BTS system into seemingly randomly scattered fragments or packets. The data packets are transmitted as cipher text encrypted by the A5 algorithm. That, plus the fact that the packets do not arrive at a BTS in sequence, makes them rather more impervious to eavesdropping but cannot be regarded as a serious security measure. Unlike the GSM security, the BTS does not decrypt the data packets and therefore is not supplied with the secret session key Kc, but simply forwards them to the serving GPRS support node (SGSN), which does decipher the packets and reassembles them before forwarding them on to the GSM backbone for delivery to the remote party. The SGSN also plays an important role in authenticating the mobile station with the HLR, rather than the MSC procedure as with GSM. In this manner, the security of encrypted data over the Um interface is extended to the location of the SGSN station. Security over this section is also improved by the fact that the GPRS uses a new A5 algorithm. However, if the HLR was arguably the most vulnerable component in the GSM system, the introduction of the SGSN station provides a new target for those wishing to attack the system.

The addition of a new A5 algorithm is no doubt an important step in improving, perhaps only temporarily, the system security, but a stronger A3 algorithm is certainly required to protect the SIM card against cloning attacks. How long the GPRS can claim the degree of security that GSM initially professed, remains to be seen. One thing that is certain is that it will be very much under the microscope and subjected to every manner of analysis as academic institutes and the like rise to the challenge. As far as the security of encryption is concerned, it is a healthy competition that inevitably will lead to the evolution of more secure communications systems. Bearing in mind the failure of GSM to provide up-to-date security, those bodies requiring high-security communications over both the GSM and GPRS networks, need to look for another solution. End-to-end encryption by strong algorithms supported by efficient, secure key management will be the best answer to the threats of the future.

6

Security in Private VHF/UHF Radio Networks

There are a multitude of ultra high frequency (300–3000 MHz) (UHF)/very high frequency (30–300 MHz) (VHF) radio networks in existence, and whilst most of them are of a military or law enforcement application, there are also any number of private networks used by dignitaries and senior political figures. As most of this book has concentrated on martial networks, the author wishes to offer VIPNET, shown in Figure 6.1 as an alternative environment. VIPNET is an example of a small network made up of a number of subnets, each with its own function and features, which need to be analysed for the implementation of security techniques. There are many possible variations of design and functions, and the addition of other components such as voting selectors might seem attractive. However, this book is more concerned about network security than the structure of radio networks, and so the author has modelled a reasonably complex system, sufficient to illustrate the principles of securing the system by cryptographic means. In fact, and as is so often the case, communications security is achieved by the implementation of a package of devices rather than a single entity. Therefore, it should be appreciated that the control of the radio medium features also makes a significant contribution to security of the network.

6.1 Applications and Features

The sub-groups are the ship group (SG), the escort group (EG), the close support group (CSG), the HQ office (HQO) and also the normal telephone subscriber group that also needs to be taken into consideration.

6.1.1 The Ship Group

The purpose of the SG is to offer a means of transportation, protection and provide a base for communications with the rest of the organisation during extended periods of travel. To satisfy this role, the ship must provide for secure telephone communications between its own PABX and radios as well as to distant centres of the organisation by means of a SATCOM link to the PSTN system. (Secure telephone systems are discussed in Chapter 4 and therefore only given cursory treatment here.) Naturally, and somewhat unfortunately for the security manager, a

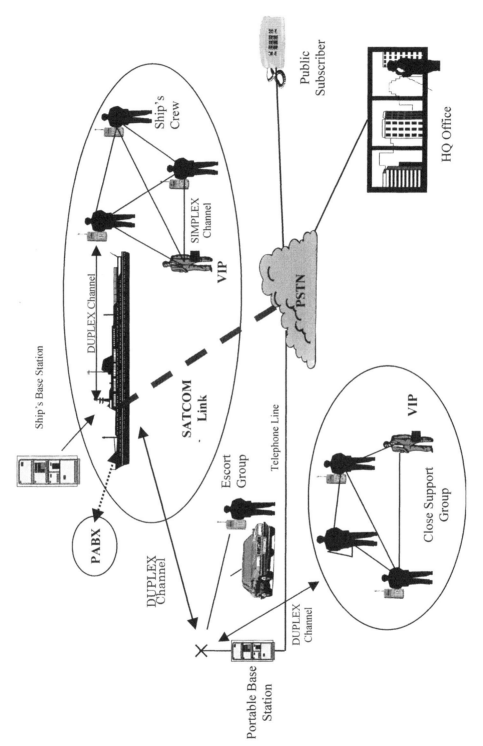

Figure 6.1   VIPNET structure

link to the outside and highly vulnerable world of the PSTN and public subscribers must be catered for.

The specification also requires for the ship to have radio communications between members of the crew, the VIP and the ship's communications centre. As will be seen in Section 6.6.2, the frequency management will allow for point-to-point radio calls between individual hand-held units, alternatively known as *sel-calls* and *group calls*, which is the simultaneous communication to all members of the group switched to the appropriate channel. This will be a *simplex* operation, and in practice, numerous channels would be made available for such purpose. To support the crew operations, the ship would be installed with a radio repeater of 15-W transmitter power, the main purpose being to extend the range of the crew's and VIP's hand-held stations. This means that some frequency changes within the channel must be made for the repeater operation and therefore communications between the mobile units and the ship's communications centre must be of a semi-duplex nature. In addition to the 15-W radio repeater for operations in the vicinity of the ship, a second radio repeater of 20 W is required to allow communications with the land-based escort group. Hence, a duplex channel must be allocated for this link.

A further requirement is that the VIP and, if permitted, other senior staff members of the organisation may wish to make telephone calls. These calls will be from a mobile hand-held unit and may be to either the HQ or some other subscriber via a PSTN provider. Provided that the unit is within the range of the ship's repeater and the correct semi-duplex channel selected, the hand-held users will be able to request that the repeater connect the call to the PSTN though a SATCOMM link. In this way, the VIP will be able to make and receive calls from any telephone in the world, including ciphered communications within the organisation's structure.

6.1.2 The Escort Group

This group has the task of transporting the VIP on land by motor vehicle, to the location desired, and the second task of effectively extending the range of the ship's communications to the shore, by carrying a portable base station. This unit would then be providing a radio communications link to the ship, its PABX to crew members or passengers and even beyond to the PSTN via the SATCOMM. To meet this demand, the EG will carry a 10-W portable repeater that will provide the link to the ship. It is envisaged that the base station would be located on an embassy roof, e.g. when such occasions arise whereby the ship to shore link is not possible due to range problems, then it will be necessary for a land-based VIP to connect to the PSTN through the EG repeater and its associated telephone line. Hence, the EG will need at least two channels, one duplex channel to link with the ship's repeater and one semi-duplex link to provide a telephone call requested by the VIP.

The remaining function of the EG repeater will be to extend the range of the CSG units, when necessary, by the semi duplex channel and allow radio sel-calls and group calls to take place between the EG, the CSG and the ship's crew.

6.1.3 The Close Support Group

As the title suggests, the purpose of the CSG is to provide a bodyguard function for the VIP whilst on visits. This subnet will require simplex sel-call and group call functions to be supported between the VIP and their support team. As radio communications are in the UHF band, the range of links is therefore limited to 'line-of-sight' operation. This mode of operation is especially susceptible to communication problems when working in a built-up area, and so to alleviate the limitations, the CSG will have their own man pack portable repeater of 5-W transmitter power. The CSG users will need an additional semi-duplex channel for this local repeater function.

6.1.4 The Telephone Groups

Broadly speaking, there are two members of this group, i.e. the HQO and public subscriber. For secure communications to and from the rest of VIPNET, the HQO will need to be provided with an encrypted telephone/s that is/are compatible with both the radio cipher units and the secure telephones on the ship's PABX. This interfacing between radio and telephone cipher units provides a technical challenge to provide a suitable solution.

However, the public subscriber communications through the PSTN are the main source of concern to the network manager. Whilst one might expect the VIP to be, at best, cautious in his 'public' conversations, the same might not be the case for other members of the organisation. Therefore, the question of access to telephone operations is a major issue.

The best that can be done as far as cryptographic security is concerned is to provide encrypted radio communications up to the point where the VIPNET signals must be converted to plain mode for the PSTN calls. This interfacing is the role of the Telephone Patch (see Section 6.5.4).

6.2 Threats

Having established the communication requirements of the VIPNET client, the next task is that of analysing the risks to the confidentiality, the integrity, the authenticity of the network's communications and the various access rights of its personnel.

These areas of threats are discussed in more detail in Chapter 1, but the application-specific threats are outlined below.

6.2.1 Confidentiality

The most obvious source of threat to VIPNET communications confidentiality is that of calls being eavesdropped over the PSTN and the radio channel, although the PABX operations on the ship should not be ignored either. There is little that can be done, as far as calls to and from the public are concerned, except as mentioned above, by encrypting the VIPNET part of the signal path up to the telephone patch. Otherwise, strictly enforced personal discipline about transmitted subject matter is required

6.2.2 Integrity

As all messages are 'real time' messages with no storage features, the possibility of an intruder modifying message data is extremely remote. The only danger would come from 'message replay', and this is more of a question for authenticity rather than the threat to data integrity.

6.2.3 Authenticity

The danger is that the identity of the other party in a call is not always apparent, and therefore, the message authenticity cannot be certain. This vulnerability can be exploited by 'message replay', whereby a third party eavesdropper can record a genuine voice message, even if it is encrypted, and then retransmit it at a later time. The eavesdropper will not be able to understand the message, as it would be ciphered, but would be able to transmit it to a receiver who would decipher the message as genuine. The real purpose of such an attack is to elicit confusion within the client's organisation.

6.2.4 Access

There are two concerns on this subject. The first is the concern of personnel, certified or otherwise, gaining access to the network, its components and communication channels. The second is that of unauthorised access to the programming of radio facilities and cryptographic parameters of the network hardware.

6.3 Countermeasures

Once the threats have been identified, the security manager of VIPNET must decide what countermeasures are available and how best they might be implemented. The cryptographic solutions are generally discussed in Chapter 2 but are more specifically addressed below.

6.3.1 Protection of Confidentiality

Communications confidentiality with the HQO will be assured by strong encryption as defined in Chapter 2. VIPNET calls for two levels of confidentiality protection: that of the entire network from external eavesdroppers and that of VIP channels from within the network.

The secrecy of all radio channels and that of radio/telephone links to the HQO through the PSTN will be protected by symmetrical encryption and the allocation of specific secret keys to each channel or group of channels. If all radio channels are so encrypted, it will not be possible for an outside party to monitor calls. Furthermore, if each channel is ciphered by a different key, it will not be possible for a member on the lower levels of the organisational hierarchy to be able to listen in to conversations of a VIP using their own secret key. The network security is strengthened by thoughtful key allocation such as assigning different keys to different groups, with common keys for use between groups as required. This improves confidentiality and provides for seamless changes on the occasions when mobile radios go missing. The danger is that if all transceivers carry all of the network keys, then the loss of one unit compromises the whole network. Sections 6.4 and 6.6.2, as well as Table 6.1, give more details about suggested key allocation.

6.3.2 Authentication

The problem of 'message replay' can be resolved by implementing *time authentication*. The effect of this is that voice calls are 'time- and date-stamped' so that any message exhibiting an

Table 6.1   VIPNET frequency and parameter assignment (abbreviated)

Channel	Transmit	Receive	Secret Key	Mode	Plain O'ride	Tel. App.	Description
1	450.000	460.000	1	S	N	N	General Hand held use-CSG
2	450.125	460.125	1	S	N	N	General Hand held use-CSG
10	451.125	461.125	1	S	N	N	General Hand held use-CSG
11	451.250	461.250	2	S-D	N	N	CSG use with 5Watt manpack repeater
12	451.375	461.375	2	S-D	N	N	CSG use with 5Watt manpack repeater
15	451.750	461.750	2	S-D	N	N	CSG use with 5Watt manpack repeater
16	451.875	461.875	3	S-D	N	N	CSG / EG use + 10Watt portable rep.
17	452.000	462.000	3	S-D	N	N	CSG / EG use + 10Watt portable rep.
19	452.250	462.250	4	S-D	Y	Y	CSG / VIP phone via 10Watt port. rep.
20	452.375	462.375	3	S-D	N	N	CSG / EG use + 10Watt portable rep.
21	452.500	462.500	5	S-D	N	N	CSG / EG to ship link via 20W port.rep.
22	452.500	462.500	5	S-D	N	N	CSG / EG to ship link via 20W port.rep.
25	452.500	462.500	5	S-D	N	N	CSG / EG to ship link via 20W port.rep.
26	453.125	463.125	5	D	N	N	Full Duplex Ship/Shore Link
27	453.125	463.125	5	D	N	N	Full Duplex Ship/Shore Link
30	453.125	463.125	5	D	N	N	Full Duplex Ship/Shore Link
31	453.750	463.750	XXX	D	N	N	Spare Full Duplex Ship/Shore Link
32	453.750	463.750	XXX	D	N	N	Spare Full Duplex Ship/Shore Link
35	453.750	463.750	XXX	D	N	N	Spare Full Duplex Ship/Shore Link
36	454.375	464.375	6	S	N	N	EG general group use
37	454.500	464.500	6	S	N	N	EG general group use
40	454.875	464.875	6	S	N	N	EG general group use
41	455.000	465.000	7	S-D	N	N	Crew + 10Watt ship repeater
42	455.125	465.125	4	S-D	Y	Y	VIP - phone via 10Watt ship repeater
45	455.500	465.500	7	S-D	N	N	Crew + 10Watt ship repeater
46	455.625	465.625	8	S	N	N	EG / CSG / Crew general use
47	455.750	465.750	8	S	N	N	EG / CSG / Crew general use
50	456.125	466.125	8	S	N	N	EG / CSG / Crew general use
51	456.250	466.250	9	S	N	N	General hand held use - Crew
52	456.375	466.375	9	S	N	N	General hand held use - Crew
60	457.375	467.375	9	S	N	N	General hand held use - Crew
61-70	457.500	467.500	XXX	D	N	N	Ship to Shore link - spare

inexplicable delay during transmission will not be received by any VIPNET station. The process may be carried out by the use of a 'time window' for reception or by influencing the synchronisation of the cipher units. This facility does not come entirely free of charge as far as complexity is concerned, as each unit will have to be programmed with the same time that must be maintained within reasonable limits. Allowing a tolerance of a few minutes will account for any difficulty of synchronising clocks.

In voice communications, the most common method of authentication is by voice recognition. With the use of vocoders, such as the AMBE vocoder, voice quality and therefore recognition is not the hit or miss affair of the past and even of some telephone systems of the present day. The fact that VIPNET is a relatively small network leads one to assume that operating personnel will be familiar with each other, and therefore, authentication by voice recognition becomes even more assured.

Cryptographically, with the efficient planning of a key structured network, secret key assignment would go to some length in assuring authentication of the calling and called parties, e.g. considering the CSG operations on a simplex channel. When communicating on a channel encrypted by a key specific to that channel, all members of the group can be assured that any call they receive will have originated from a genuine unit of the group. Similarly, those initiating a call can be certain that only those units possessing the correct secret key will receive their message. However, within the CSG, no member can be sure, in cryptographic terms, of whom in the group call was the source of the message. The digital signalling of radios allows the transmission of the unit ID, and provided that the hand-held ID is genuine, the reception and display of this ID by a receiving party is usually accepted as an authentication of that caller.

The use of selective calling (sel-call) enhances the authenticity of a resulting conversation. The sel-call process demands that a calling unit accesses an appropriate channel and enters the ID of the intended receiver before pressing PTT. All hand-held radios in the group will normally be in the standby mode and therefore will be scanning the channel in use, but only the radio with the correct ID will be able to receive the message and hence impart confidentiality on a point-to-point link.

A small network such as VIPNET would normally have enough channels available to allow considerable opportunity and flexibility in the frequency management permitting the use of more channels for point-to-point operations. See Section 6.6.2 for more details on the frequency/key management.

Extrapolating from this argument, it can be seen that authenticity within the whole network can be achieved by cryptographic key management and voice recognition. The fact that a radio may be stolen underlines the role and value of voice recognition as a useful, if not essential, component in the authentication process.

6.3.3 Access Control

There are two aspects to consider as far as access is concerned:

* Access to radio/hand-held operations
* Access to the programming of:

    a) Radio medium functions including channel access
    b) Cryptographic parameters

Access control in the first instance is affected by the implementation of an operator pass-word. This will give the user, having the knowledge of such information, the right to use that particular transceiver and whatever features are pre-programmed within. However, there should be no possibility for the operator or third party to be able to modify the radio and security parameters programmed by a higher authority, the network manager.

Programming access within a small network such as VIPNET would probably only involve one person and perhaps a supporting assistant. In large networks, it is likely that two different management figures would be needed. One body would be needed to manage the radio and communication aspects of the network, i.e. the *network manager* and a second, the *security manager* to control the cryptographic parameters.

The network manager would be responsible for:

- The frequency assignment and channel designation including the simplex and duplex operating modes
- The programming of radio operational features such as sel-call set up, data messages, etc.
- The programming of macro software programs, i.e. radio subroutines that control emer-gency clearing, alarm calls and their elicited responses, display features, blocking and parameter cloning between radio units, etc.

It is highly likely that the network manager will wish to create a hierarchy of users, through operational feature management. On the one hand, a basic user who might not be so trust-worthy might simply be given the control of transmitting voice on one channel only, with no rights to change channel, make sel-calls or telephone calls. On the other hand, the VIP will expect to have all modes of communication available to them but would not wish to be burdened by the technicalities of either the radio parameters or the security operations. The VIP would wish for 'operating transparency'. In between these extremes, there will be operators and senior staff requiring a range of features to be able to carry out their daily functions.

The security manager will be responsible for:

- Developing a security strategy for the network as a whole
- Constructing a security hierarchy
- Key management including

  Key generation
  Key distribution
  Key destruction

- Password management
- Operator training in co-operation with the network manager

Access to the security manager's menu must be strictly controlled by a password, as any unwarranted modification to any of these parameters would increase the vulnerability of the network to attack. So, for VIPNET, a two-password access control, i.e. an operator and security manager, would be sufficient, but for larger networks, three levels would be necessary.

6.4 Communications Network Design and Architecture

This section looks at the design and interoperations of the subgroups and their components.

6.4.1 The Close Support Group

The purpose of the CSG is to give protection to the VIP during visits ashore. To carry out this function, its members will need to communicate with each other and with the VIP, and on occasions, it is envisaged that the team will communicate through the ship's communications centre.

The group consists of three guards and the VIP, as shown in Figure 6.2. Group calls, which would form the majority of traffic, will be by simplex operation on a range of channels represented by CH.1–10. When communications become difficult, e.g. due to operations in a built-up area or where radio range is questionable, it will be possible for the team to switch to a semi duplex mode and engage the group's man-pack repeater by selecting the appropriate channel, i.e. CH.11–15 on the hand-held units. The man-pack repeater would be carried covertly in an ordinary shoulder bag. By selecting channel CH.16–20, another semi-duplex channel group, any member of the CSG will be able to access the EG repeater when within range of it and subsequently communicate with the ship (see Figure 6.3). The VIP and perhaps a senior member of the CSG will be able to access the telephone channel provided by the EG repeater by transmitting to it on channel CH.19. The basic CSG team members will not be able to use this channel, and it will not be programmed into their mobile radios. Neither will be the corresponding secret key SK.4 (see Table 6.1).

For telephone calls from the PSTN to the VIP mobile radio, the calling parts would simply need to dial the telephone number of the repeater location plus an extension derived from the transceiver ID to make a point-to-point connection with the VIP.

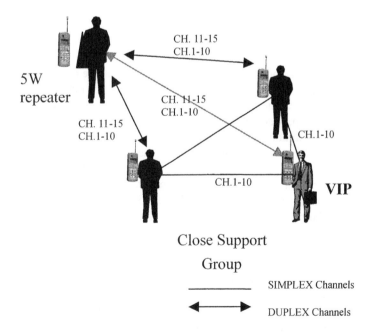

Figure 6.2 The CSG radio operations

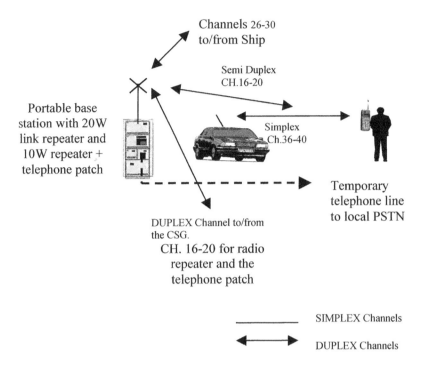

Channels 26-30
to/from Ship

Semi Duplex
CH.16-20

Portable base
station with 20W
link repeater and
10W repeater +
telephone patch

Simplex
Ch.36-40

Temporary
telephone line
to local PSTN

DUPLEX Channel to/from
the CSG.
CH. 16-20 for radio
repeater and the
telephone patch

——————— SIMPLEX Channels

◀——————▶ DUPLEX Channels

Figure 6.3   The EG radio functions

6.4.2 The Escort Group

The EG consists of one or two vehicles with drivers and a portable base station consisting of a 20-W radio repeater and a second repeater of 10 W with a telephone patch function capable of connecting to any available local PSTN number. The purpose of the 10-Watt repeater is to extend the range of the CSG and its own team members and to provide a link between these two groups and to the ship. Members of the CSG and EG will be able to communicate with each other either by a common Simplex channel CH.1–10, or through the repeater channels 16–20. The 10-W repeater will also provide a link to the local PSTN via the telephone patch for any user with access rights to channel CH.19. Users within the EG will normally communicate with each other by a range of simplex channels, labelled CH.36–40. Details of the repeater operation can be found in Sections 6.5.2 and 6.5.3. All mobile radios of VIPNET are able to access the radio repeater by transmitting on channel CH.16–20.

6.4.3 The Ship Group

The ship performs the role of a temporary base for the VIP during extended visits to regions distant from the HQ Office (see Figures 6.4 and 6.5). The group consists of the crew and VIP and carries the main base station comprising a link repeater of 20-W and a ship group repeater of 10-W output power. The purpose of the link repeater is to provide an extended range for the crew and the land-based mobile units, e.g. the EG and CSG. Access to the repeater, when in range, is achieved by selecting channels CH.31–35 (see Table 6.1). Both the EG and CSG will

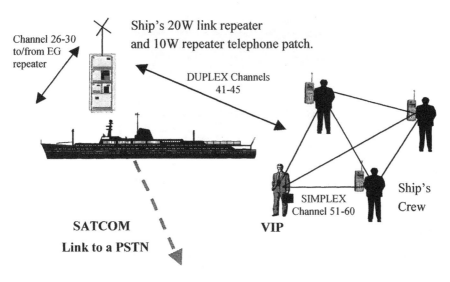

Figure 6.4   The ship group radio functions

be able to access the ship's 10-W repeater by selecting channels 41–45. The crew normally uses the simplex channel CH.51–60 for selective calls and group calls within the group.

The ship is installed with her own PABX system comprising secure telephones throughout. It is linked to the PSTN system by a SATCOM transceiver. Accredited users, such as the VIP transmitting on channel CH.42 can access the controlling telephone patch and make calls to subscribers via the PSTN and to the PABX terminals.

The ship's repeater is linked to the land-based mobile repeater through channel CH.26–30

6.5 Hardware Components and Functions

The section gives a brief insight into the actual construction and operational characteristics of the main network components and how they interact to produce an efficient and secure network.

The major components of VIPNET are:

- Hand-held UHF radios
- Manpack repeater
- Base stations/repeaters
- Telephone patches
- Security management tools

6.5.1 Hand-held UHF Radios

The hand-held radio units are UHF transceivers, operating in the 450–475-MHz band, being encrypted for two-way communications, with the radio coverage range being identical, or very similar, for both plain and ciphered operation.

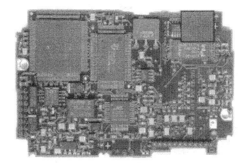

Figure 6.5   An encryption device integrated into a hand-held transceiver circuit board

Required features:

- Pre-programmable function keys for customer specific applications
- Status indication (ciphered/plain) by coloured LEDs
- Lockable numeric keyboard
- 1000-channel capacity
- Eight channel scanning programs (searches up to 12 channels for a transmitted signal)
- Various scan detection mediums, i.e. carrier, sensitivity, tone, DTMF, sel-call, squelch
- 20 predefined messages of 18 characters length each
- Call-back memory of last eight messages used
- Quick-dial memory for 10 DTMF telephone numbers
- Set-up parameter programming via keyboard
- Access restrictions by password hierarchy
- Parameter programming software package for macros

Required security features:

- Excellent voice quality for voice recognition
- Independent operating groups possible
- Interoperability of groups
- Late entry capability (continuous synch of cipher module)
- Default key operation (programmable option)
- Plain override (programmable option)
- Warning alarms for plain operation, loss of keys
- Built-in test (BITE) facility

The block diagram of a typical hand-held radio transceiver is illustrated in Figure 6.6. High-quality mobile radio units should have similar features to the following technical specification:

Technical specification

Frequency range: 450–475 MHz
AF bandwidth: 300–2600 Hz
Operation mode: simplex and semi-duplex

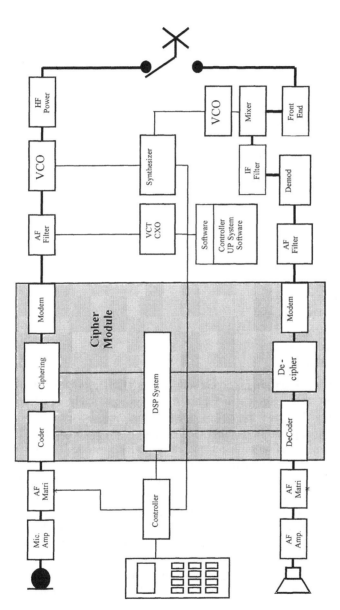

Figure 6.6 Block diagram of hand-held transceiver

Number of channels: 1000 max.
Modulation: PM/FM
Battery: 7.5 V
Capacity: 700 mA at 10%, 10%, 80% (RX, TX standby)

Transmitter characteristics:

Radio frequency (RF) output power: 0.1–2.5 W
Adjacent channel power: < -6 dBc (12.5-kHz channel spacing)
Signal-to-noise ratio: >40 dB

Receiver characteristics:

Sensitivity: 0.45 (V/50 Ohm/12 dB SINAD
Signal-to-noise ratio (AF): 38 dB
AF distortion: $<7\%$ at 1 kHz, deviation 60% max.

Cipher unit:

Operating mode: semi-duplex
Transmission mode: binary NRZ, FSK
Data rate: 4.8 kb/s
AF bandwidth: 300–3000 Hz
Synchronisation: continuous
Secret keys: variety 10^{38}
Secret key capacity: 3*10
Transfer key: 1*32 characters
Key period: >180 years
Algorithm: prefer proprietary
Initialisation: random

For more covert operations, transceivers such as the VECTOR (PRM5202A) from Thales are ideal. The 'VECTOR', shown in Figure 6.7, is an extremely compact radio, which, although specifically designed for covert applications, may be used in an overt role in conjunction with the appropriate ancillaries. The radio is rugged and splash proof, and is intended primarily for use by surveillance operators (police, customs and excise, drugs squad, government agencies and body guard nets, etc.).

The radio provides continuous operation over the 380–430-MHz band and supports TETRA DMO & TMO (direct & trunk mode operation). RF power output is 1 W when battery-powered, and 3 W when mounted in a vehicle. The end-to-end encryption is provided by a removable module. This module is protected by a comprehensive anti-tamper system that erases stored keys and algorithms if attempts are made to extract them. Traffic keys may be filled via the infrastructure using over-the-air re-keying (OTAR), and remote control of many programmable features is possible (OTAC). The radio may be operated in clear or air interface encryption-only modes to enable communications with non-secure TETRA users.

Both secure and non-secure peripheral equipment interfaces (PEI) are provided for connection of external serial data devices. VECTOR has no external operator controls, except for a key zeroise or emergency clear. A range of ancillaries allow the radio to meet both covert and

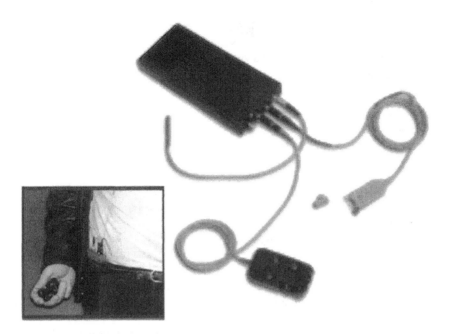

Figure 6.7   The vector covert transceiver from Thales (courtesy of Thales Defence Ltd)

overt roles. Discreet operation is provided by a radio control unit (RCU) that is designed to fit unobtrusively in the palm of the hand and can be operated by touch alone. For overt operation, the control loudspeaker unit (CLU) provides control facilities integrated with a loudspeaker/microphone in a convenient clip-on unit. The mobile radio allows for discreet installations in cars and motorcycles.

The radio control software is contained in FLASH-programmable memory, which can be readily updated to utilize future TETRA system enhancements. The main case contains the radio and cryptographic components, packaged in a compact, slim, lightweight unit, which is ideal for carrying unobtrusively about the person.

Consisting of only two PCBs mounted in a simple two-part clam-shell case, access for repair is extremely easy. The casework design is light and strong, with a minimum number of fixings. The radio weighs less than 300 g and has a volume of less than 120 cm^2.

Technical specifications

General:

  Frequency range: 380–430 MHz
  RF carrier spacing: 25 kHz, in 6.25-kHz steps
  Power output (with standard or extended-life battery): +30 dBm average (TETRA Class 4)
  Mobile power output: +35 dBm average (TETRA Class 3)
  Radio channel: half duplex (audio full duplex to PSTN)
  Operating modes: TETRA TMO, TETRA DMO
  Air interface encryption: TETRA air interface encryption using TEA 2

End-to-end encryption:

Internal encryption module can support a variety of encryption algorithms, including certified UK enhanced grade:

   Battery (standard life):
   Rechargeable Li ion, 7.2 V nominal, 9 h life in 1:1:18 duty cycle
   Weight: 300 g (excluding battery)
   Dimensions: 65 × 105 × 12 mm

External connectors:

   Antenna: 50-Ohm SMA coaxial socket
   Audio 1: 7-way socket
   Audio 2: 7-way socket
   Peripheral equipment interface: 5-way socket

Controls and indicators:

   Radio: the only control on the radio is the crypto-zeroise button.
   Earpiece: in-ear inductively coupled covert earpiece. Ear-hanger earpiece (connects to CLU)
   Control loudspeaker unit (CLU): LCD dot matrix display 47 × 25 mm, 80 × 32 dots, 4 menu selection buttons, numeric keypad, speaker + button, speaker – button and PTT.
   Radio control unit (RCU): 10 position rotary switch: on–off/talk group/function select step up & down push buttons.
   PTT button initiates normal transmission of microphone signal.
   Single tone push button initiates transmission of an audible signalling tone, replacing the microphone signal.

Ancillaries:

   Antennas: a range of overt and covert antennas is available, depending on the operating frequency and role.
   Charger: battery charger for the above batteries.
   Harness and pouches: shoulder harness, leg pouch, waist pouch and vest harnesses.
   Control units: control loudspeaker unit (CLU) (overt) incorporating microphone, radio control unit (RCU) (covert)
   Microphones: covert microphone incorporating inductive loop.

6.5.2 Base Stations/Repeaters

There are two base stations: one located in the ship operating as a base station and one as a portable unit with the Escort Group (see Figure 6.8). Each has almost identical functions and hardware structure except for differing output powers. Only the ship's base station repeaters are shown here. Neither of the radio repeaters allows monitoring of the ciphered calls. The function of the base stations is to provide:

   Ship-to-shore communications over a full duplex link

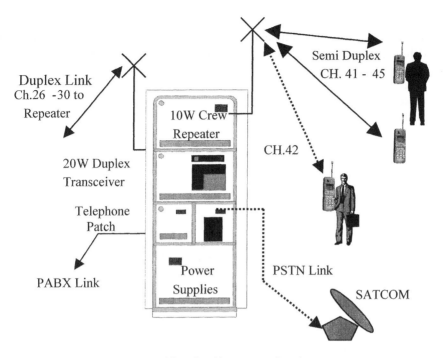

Figure 6.8   The ship repeater functions

A repeater facility to extend the range of the hand-held radios operating within the group or region
A link between a radio and the PSTN or ship's PABX telephones and vice versa via a semi duplex link

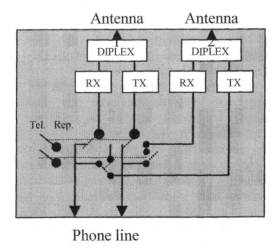

Figure 6.9   Repeater switching between radio transceivers and telephone line

Figure 6.9 shows how the repeater switch between radio repeater and radio to telephone line is arranged. A signal from a mobile radio on channel 42 arriving at Antenna 2 will be switched through to the transmitter of Antenna 1 for transmission to another mobile radio via the land-based repeater station. A similar call made on channel 41 to Antenna 1 will result in the call being repeated to another mobile linked to that repeater. The final option using the ship repeater is for a hand-held mobile to make a call to a telephone via Antenna 1 (from the EG repeater), and then the ship's 20-W repeater will switch the radio call via the telephone patch to the dialled telephone location. The actual switchover in the latter case depends upon the calling criteria of a mobile radio that will be either a radio ID or a telephone number. Calls to a radio that are originating from a telephone will switch the phone line to the relevant base station transceiver.

6.5.3 Telephone Patch

The telephone patch is a transcoder unit that is used to convert signals between the radio cipher module and the telephone cipher module and provide a link between the two mediums. Two different modules are required as the synchronisation processes of each are quite different. The radio cipher synch depends largely upon the vagaries of the radio medium such as multi-path fading caused by the differing topographies. From Figure 6.10, it can be seen that the patch consists of:

- A radio cipher module
- A telephone cipher module
- An interface unit

On receiving a call from a radio, requesting connection to a PSTN telephone line, the repeater will pass the radio signalling to the interface where the digital 'dialling code' is converted to dual tone multiple frequency (DTMF) signals. The DTMF signals are passed to the telephone cipher unit, which allows the dialling to take place in plain mode. The digital dialling code from the radio is converted into DTMF tones, which are then passed onto the main signal line in the interface module, between the radio and telephone cipher units. The DTMF signal is then passed to the PSTN via the SATCOMM, to complete the call. On successful connection to the called telephone, the conversation can be continued in plain or ciphered mode, depending upon the security manager's strategy and also the availability of a secure telephone at the PSTN end of the link. Either the radio or the telephone user should be able to make the change from plain to cipher mode, but one would expect that both parties would be required, in agreement, to control the change from cipher mode to plain mode. Suitable alarm signals, audio and visual, would be employed to notify the users of the transmission status. In cipher mode, the voice signal from the radio will be ciphered at the hand-held radio, then, on reception by the repeater, deciphered by the radio cipher module and then re-ciphered by the telephone cipher module before being transmitted to the PSTN number. An 'on hook' condition or radio channel closure would close the link.

When the reverse operation takes place, i.e. a PSTN telephone calls a hand-held radio station, the caller will dial the location of the repeater in question. This will be on a SATCOMM number in the case of the ship and the location phone number designated to the land-based EG portable repeater. The plain DTMF signals from the PSTN will be received by the telephone patch interface, which will convert them into the digital codes

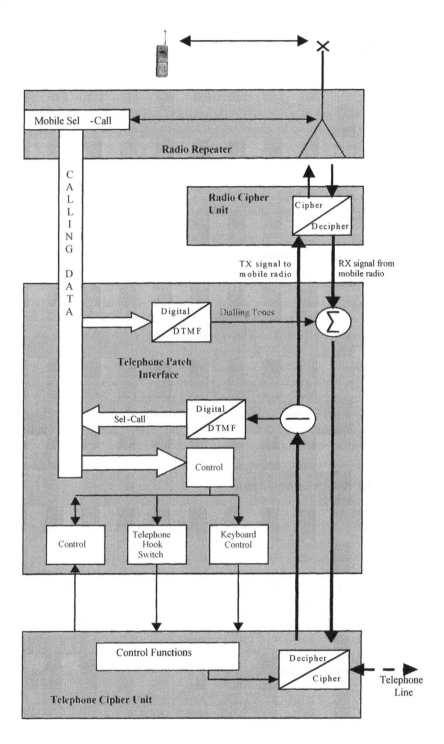

Figure 6.10 The basic functions of the telephone patch

of the radio signalling. The repeater then makes what is essentially a selective call to the radio required by the caller, and once the link is established, the change from plain to cipher mode can be made. This is assuming that the calling parties have the freedom to make such a decision.

The fact that a radio/telephone cipher conversion is required provides a slight opportunity for the exploitation of a plain voice signal being available in the telephone patch circuit. One would not want such a situation to exist in an unattended base station or repeater, unless further protective measures were taken. As far as the radio repeater operation is concerned in the VIPNET, there is no deciphering in the base station and therefore no threat of eavesdropping at that point.

6.5.4 Security Management Tools

The security management tools will come in two parts to match their functions. The main management centre will be a computer with secure access, key generation and storage facilities. It will be treated in a highly confidential manner and will be accessible only to the security manager and assistant. The second component will be the courier device, which, as the name suggests, will provide a method of distributing keys, security parameters and the radio programming of access rights and frequency assignment. Military networks would usually require a more robust courier device, and these are often referred to as 'Key Fill Guns'.

In a small network like VIPNET, it would be most practical to use an encrypted laptop computer as the management centre and a second laptop as both a back-up and courier device. Bearing in mind the possible geographic dispersion of the network members, a useful addition would be the inclusion of chip cards as a distribution aid. If the network manager has a number of distribution options available to cater for the ever-changing and unpredictable circumstances, the operational continuity should not be too difficult to maintain.

6.6 Security and Key Management

This section discusses the implementation of security features into the VIPNET application. It looks at the frequency management, security administration and key/operating parameter management.

The management centre would be located on board the ship and would be used to generate keys, user access rights, frequency assignment for the radios and as an archive for these parameters. Relevant data will be downloaded into the courier laptop computer for the direct distribution of data to the mobile radios by means of a connecting cable. The option of a chip card programmer and reader would allow for the secure distribution of data to distant locations such as the HQ office and each secured telephone that might be added to the network. The use of chip cards as a distributing medium was discussed in Chapter 2, and, as suggested, the cards must be protected by the encryption of their contents with a key transport key (KTK).

6.6.1 Functions of the Management Centre

These should include:

- Key generation from a built-in true random number generator (probably best located in an attached PCMCIA standard module)
- Operating parameter definitions
- Secret key to channel assignment
- Time authentication programming
- Key and parameter distribution
- Key and parameter loading to/from chip card devices
- Verification of downloaded key data
- Secure storage of sensitive data

The laptop computers will have two access levels, i.e. the courier access and the security manager access, and should incorporate the following security features:

- Failed password login attempts should, after three failures, result in the destruction of key data in the courier devices
- A read-out limit should be imposed on courier devices so that an excessive number of readout attempts cannot be made
- Attempts to gain physical access to the hardware interior should result in the destruction of sensitive data
- Provision for emergency clearing of contents
- Verification checks available to monitor the IDs and contents of all secure components
- A log of events so that efficient and secure operation can be monitored

6.6.2 Frequency Management

Hand-held radios normally operate in either simplex mode, for direct sel-call, or group calls to other hand-held units. In this case, no other components are used in the radio path. For use with the repeaters, the mobile units must communicate with the repeater in half duplex mode, whilst the links between the repeaters are full duplex channels, i.e. they transmit and receive simultaneously over the same antenna. The radios are programmed within the desired bandwidth of 450–470 MHz with a channel spacing of 12.5 kHz. Normally, a quality radio is capable of being programmed to operate with up to 1000 channels, but VIPNET does not require so many. The duplex spacing between the transmitting and receiving frequencies of a channel is 10 MHz.

The VIPNET model is shown with abbreviated channels, and it is normally the case that many more channels would be available for use. The CSG, for example, may have some 50 channels to select from, depending upon what other radio traffic is being carried by those frequencies. However, the key/channel allocation would probably be the same for each channel in the sub-group. With efficient management, the security operations should be transparent to the user, thus preventing any operating mistakes or illegal influence of the set-up. All the radio operator needs to do is to select the pre-arranged channel on which to operate, and the correct secret key and operating parameters will be automatically assigned.

As there are only two repeaters in each of the base stations, one for the ship to shore link and one for the group communications, a special sequence of signalling is required by those users wishing to access the telephone patch. This signalling will cause the relevant repeater to switch to telephone patch operation instead of acting as a radio repeater, and so, simultaneous radio repeater and patch operations would not be possible. Should either radio or telephone traffic become excessive, a dedicated repeater would need to be added to cover the telephone communications. The key to channel frequency and parameter assignment is shown in Table 6.1.

Table 6.1 shows the typical start position of a security set-up strategy often referred to as the 'deny all' position. The term comes from the fact that all users, except the VIP, are denied all options to operate in plain mode and do not have telephone access. There is little doubt that this strategy is the best for the initial VIPNET, but it is equally certain that it will be too rigid a structure to survive as it is, for long. One can imagine that the first modification would be to grant wider access to the telephone operation. This would most likely mean that senior staff members, e.g. the ship's captain would be given this facility as he may well need to speak to the VIP when he was located in his HQO and perhaps out of direct radio range. It is even more likely that the captain would wish to make calls to plain telephone stations via the PSTN or by radio to, say, various port authorities. Therefore, it can be seen that modifications to the parameters will vary as time passes by.

Plain override is the facility of one radio transmitting in plain mode, switching the operating mode of a receiving station, also to the plain mode. This facility does bring into question the confidentiality and authenticity of the network. It is far better to ensure that all communications are forced to use cipher mode, wherever possible, as would be the case in a closed network. The only legitimate reason for allowing 'plain override' is in the case of emergency when a VIPNET unit has lost its keys. The other case would mean that a radio external to the secure network is making the call. There may be a good reason for such a call, but it must be investigated thoroughly before allowing connection and conversation to continue.

The simplest way to cope with the incidence of lost keys is to employ a default key. Very often, the manufacturers of ciphering equipment include a default key, which usually has a permanent value, to give protection even under the worst-case scenario of 'lost keys'. A second scenario regarding lost keys is in the case of a lost or stolen hand-held radio. In this case, the default keys are of no use, and the only solution to the problem is a quick update of new keys throughout the affected system. As mentioned elsewhere, transmissions using default keys are certainly preferable to communicating in plain, but there is always the danger that too much reliance is put on the strength of this key. The fact that it is a default key and a somewhat permanent fixture and is possibly installed in other networks utilising the same product means that its security must be regarded as being 'suspect' in the least. Bearing these facts in mind, it is up to the security manager to assess the requirement for the use of default keys and either allow their use or forbid it. One can imagine that both the senior staff and the VIP would appreciate the protection in emergencies and that they could be trusted to use a default key sparingly. Basic users might not be thought of with the same confidence and therefore should be denied the option.

As mentioned previously, the user operation should be as transparent as possible, and they should not have any control over the use of secret keys. There is an obvious requirement for users to be able to change channels from time to time, according to what type of call is to be made and to whom. The channel selection process provides a means for the security manager to assign keys to frequencies rather than let individuals make a choice. For example, the CSG team are allowed to select any channel from 1 to 10 for simplex calls, these calls being protected by SK 01. Any other mobile radio from outside the CSG but within VIPNET must have both the channels and the corresponding key SK 01. In the case of channels 16—20, where the CGS and EG are using the portable 10-W repeater, the VIP has the same channels programmed in his transceiver and has SK 03, but they are the sole user of key SK 04, whereas the remaining members of the group can only use key 03. It is the possession of SK 04 on Ch.19, in this case, that allows only the VIP to access the telephone patch and beyond. The VIP does not really care about how the key selection is achieved, but simply needs to know that, when he selects channel 19 or, for that matter, channel 42. When doing so, the VIP will gain access to the telephone operations and be secure in the knowledge that their calls to the dialled telephone, presumed to be a secured phone holding SK 04, will be confidential. Even if another radio can access CH 42, its user will not be able to eavesdrop, as they will not have key 04. This method can be expanded upon so that the VIP can use a selection of keys, by means of channel selection, for conversations to secure telephones having keys within that range. From Table 6.1, it can be seen that several channels are grouped together in function, but also they have the same frequencies programmed for each channel. The purpose of this duplication is not always to allow a change of frequency by changing channel, but rather to stay on the same frequencies and change the operating parameters, or keys for that channel.

Having constructed the assignment table, the next question before the security manager, is just exactly which transceivers will be programmed with which channels and, therefore, which operating features. In such a small network such as VIPNET, with probably about 20 operatives in total, one would expect a more or less common operational policy to be adopted. The subgroups of EG, CGS and the crew will expect to be able to communicate across group margins but, at the same time, to be able to converse within their own groups. They should be able to do this without fear of interference due to the heavy radio traffic generated by other groups. In this case, the different channels are distinguished not only by frequency but also by key. As the network grows and perhaps other subgroups are formed, the network structure will also change. Groups will become more polarised, individual tasks and functions will change and the addition of a second or third VIP would have serious consequences for the network manager. From a relatively simple structure summarised in Table 6.1, VIPNET might well become a complex and difficult beast to manage. The most important management strategy is that of *starting simple* and letting the network evolve to meet the demands of the client.

6.6.3 Key Management

As the network is a mobile one with individuals and groups possibly not having the same central location, or at least infrequent visits to it, the best option for a key system would be the three-key system described in Chapter 2. The capacity of the VIPNET cipher units is 20 secret keys, which would be currently used at any point in time so the three-key system requires the storage of up to 60 keys. As said, 20 keys would be currently active in the *current memory*, 20 keys would have expired but will be held in *past memory*, and 20 keys should be available for the next period in *future memory*, anticipating a key change. As illustrated in Chapter 2, the system largely overcomes the problems of non-synchronized key changes, which can render a network insecure for some period, until all stations have been updated. The method relies upon individual transceivers making periodic visits to a central or divisional office at which time they may be either updated with new keys or exchanged for units already loaded with the new key data, i.e. something of an ad hoc affair.

Key changes can be initiated by a time factor loaded in to the units by the security manager, a manual action through the key board, or a simultaneous change command sent 'over the air' (OTA) from the manager. All these methods have their benefits and their limitations, and inevitably, each will fail from time to time. One option, which is most suitable for this type of network, is a 'cascade change' (see Figure 6.11).

The 'cascade' change relies upon the fact that all units within the network will have a *present key* and a *next key* stored in their key memories and that within a reasonable amount of time, all stations will be on air and communicating with each other. The actual change is initiated by the security manager, having previously loaded a new key set into the *next memory* of their own transceiver and then manually performing a key change. The next transmission the security manager makes to a fellow station, unit A, involves the inevitable handshake that carries out a key check to establish whether or not the two stations do in fact have a common key to cipher the communications. Normally, of course, the two units would find a common key within their *present memories* and, on doing so, permit ciphered communications to take place. However, as the manger's unit has carried out a key change, a common key will be found in unit A, but not within the *present key memory,* as expected. Now, the fact that the common key has been found in unit A's *next memory* is an indication to that unit that a key change has taken place in the transmitting station, and therefore, A must carry out its own key change to keep in step. Consequently, A will automatically move its *next key* set into the *present memory*, the old current key set will move to the *past memory*, and the old past key set will be overwritten and therefore lost entirely. Subsequently, both station A and the security manager's radio will make further transmissions, and the process will be repeated throughout the network until all transceivers have been updated.

Once the key change has been completed, the security manager will give instructions that users should make an effort, at the earliest convenient opportunity, to return to HQ, the ship, or other operations centre where a completely new set of keys will be waiting to be downloaded into the *next memory* from the laptop courier or from a chip card reader station. In this manner, over a controlled time frame, all radio units will receive a new key update with the newly generated key set being loaded into individual *next memory* locations to await the next key change.

Almost certainly, some station will miss an update and therefore will be unable to make a key change as required by the cascade procedure. In fact, any transceiver not having a *next key* set in its memory will be forbidden to make a change at that time, for to do so would render the three-key system untenable, as, according to the system rules, a radio may only use *current* keys for transmission. Therefore, any unit having no keys in its *present memory* would not be able to transmit in cipher mode. To prevent this highly undesirable situation arising, as soon as a key change has taken place, the unit's user will be made aware of the fact that their radio has an empty *next memory* location, and their operational discipline should ensure that they seek an update at the earliest convenient opportunity. Unless the user receives a warning that the *next memory* is empty, they will have no influence, or idea, that a key change has actually taken place.

If the cascade procedure fails for any reason, individual transceivers can be updated from a courier device as the need arises. However, what is essential to note is that should any station miss two key set changes, it would not be able to operate with the other stations. If such a situation were to arise, the offending unit will have retained a transmission key that

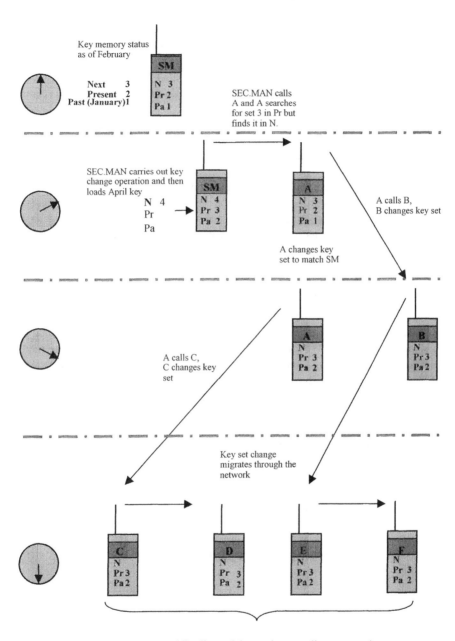

Figure 6.11 'Cascade' distribution

cannot be found in the key memories of its partner stations. A complete update from a courier device would be the remedy. Of course, if default key operation is permitted, it may be used as a final option to preserve the security of communications.

Plain override is the situation whereby one transceiver has the ability and permission, to change the operating state of its calling partner. Ideally, all communications should be in cipher mode, and only in extreme circumstances should either plain mode or plain override be permitted. As suggested earlier, such facilities like these should be denied until a specific and justified requirement is established.

Time authentication is a very useful security tool, but it must be strictly controlled in the sense that all network stations must be programmed to operate in this mode, and all stations must keep an accurate clock. A failure to implement either of these points will result in an inoperable radio unit.

Key updates to the secure telephone stations within the network will be achieved most efficiently by the use of chip cards dispatched from the key management centre in good time, to accommodate trouble free changes and the uploading of new *future keys*. Once again, the three-key system comes into its own here, as it grants some degree of flexibility over the timing of key changes.

6.7 Other Security Features

Each individual manufacturer is on the lookout to exploit niches within the industry segment, and as a result, there are many extra facilities that are added from product to product, to gain some advantage over rival companies. Most of these features are enabled by the industry's move to digital signalling, which has very much broadened the horizon of radio operations. Three of the more common features are highlighted in this section.

6.7.1 Remote Key Cancelling

This is an application that can be implemented to remove some of the dangers resulting from loss or theft of a transceiver. When a transceiver has gone missing, the security manager has the capability to transmit a command to the missing unit, to invoke what is, in effect, an emergency clear operation. Thereafter, the offending transceiver will be unable to join in ciphered communications. The ID of the lost device should be added to a black list, and any legitimate user receiving a plain call, or plain override call, from a unit whose ID should be transmitted as part of the authentication process, will be aware of the situation and take the necessary steps to avoid communications with that station.

6.7.2 Remote Blocking

This function is very similar to the above key cancelling, but rather than just bringing about the key destruction in the lost transceiver, the signal from the manager's radio unit will totally block any transmission or reception by the unit in question.

6.7.3 Silent Mode Tracking

This is perhaps the ultimate secret weapon in the security manager's bag of security tools. On the report of a transceiver going missing, the security manager can transmit a command from their own radio, to bring the offending station into the transmit mode in such a way that the new possessor is not aware of it. Then, with suitable radiolocation equipment, the missing radio can be sought for and recovered. Naturally, the search depends upon the integrity of the power charge of the battery of the unit in question.

7

Electronic Protection Measures – Frequency Hopping

The analysis of radio communications data and traffic is, without doubt, a profitable activity in any attempt to gain ascendancy over an adversary. However, as seen from previous chapters, information is not always easy to come by and this is especially true where communications are secured by encryption. When message data is well protected, then alternative methods of attacking communications are adopted. One such method is that of communications jamming, i.e. the practice of denying the use of the transmission medium to the opponent. Even this method of attack is not as easy to effectively accomplish as might, at first might seem and there are techniques available to help maintain communications that are subject to this electronic warfare. Electronic warfare (EW) is defined as all of those measures, active and passive, aimed at receiving, analysing and jamming the radio communications of an opposing force. The countermeasures are known, collectively, as electronic protection measures (EPM), formerly referred to as electronic counter countermeasures (ECCM) and this is the subject matter of this chapter.

EW can be broken down into three areas of interest, i.e.

1. Electronic support measures (ESM)
2. Electronic attack (EA) formerly known as electronic countermeasures (ECM)
3. EPM formerly known as ECCM

7.1 ESM

This area includes:

- Searching
- Interception
- Location
- Identification of communications and their sources

7.2 EA

Includes two modes:

- Passive EA
  - Such as chaff (Figure 7.1)
- Active EA
  - Jamming of communications

    Spot or narrowband
    Swept
    Comb
    Sequential
    Broadband
    Pulse

  - Deception

Figure 7.1   Electronic chaff

7.3 EPM

EPM is the name given to those measures taken to counter the threats of ESM and EA. Frequency hopping is one such countermeasure and is the main subject of this chapter.

7.3.1 Methods of Attack

As introduced in Chapter 1, normal radio communications are prone to four methods of attack:

- Interception
- Deception
- Direction finding
- Jamming

The nature of the radio communications medium ensures that whatever is transmitted, is available for all ears that may wish to listen. Therefore, the best tactic to maintain secrecy is not to transmit at all and whilst there are occasions when this approach is not only desirable but indeed essential, the enormous demand for prompt information and command, ensures that radio communications is both necessary and the most efficient medium of communication.

The degrees of communications interception, however, can be limited by the subtle use of transmit power control. There is a general belief that high transmitting power is 'all important' in getting the message through, yet there is the opportunity whereby a low power transmission particularly in the VHF and UHF bands, is just sufficient to communicate with an allied party, but out of range of opposing forces. There are transmitters on the market, which include the feature of *adaptive power transmission.*In this case, the bit error rate (BER) of a radio link is measured and the transmitter power is automatically reduced until the BER is about to fall below an acceptable level. The transmitter power is then stabilised at its optimum level for that link.

The steerable antennae or focussed radio beams, as used in microwave links can be useful against interception but have limited use in general broadcasting and do remain subject to jamming and direction finding.

Deception or 'spoofing' as it is commonly called is where the enemy or other party takes on the guise of being a friendly element. In plain voice communications this is not a difficult task and has often been used in the past to confuse the opposing forces particularly in the Second World War and more especially in air warfare. Some method of authentication is required to combat such attacks and implementing good encryption with the possibility of time authentication is the best solution.

Direction finding is the technique of locating and identifying a transmitting station for the probable purpose of future attacks on those centres. Traffic analysis coupled with direction finding would reveal much about a communications centre, making it vulnerable to both physical assault and jamming. The likelihood of detection is reduced by power minimising (see Figure 7.2), directional antenna steering (see Figure 7.3) and by adopting what are known

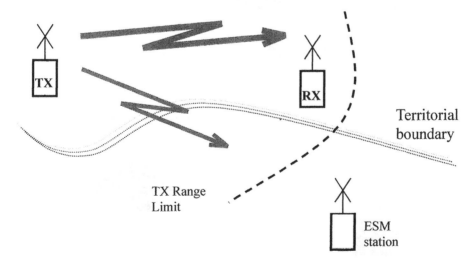

Figure 7.2   TX power minimising

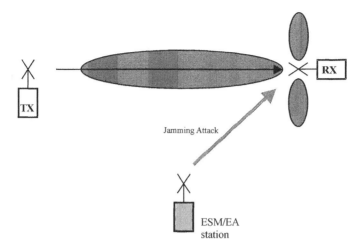

Figure 7.3   Antenna null steering

as 'spread spectrum' techniques. Spread spectrum is a method of increasing the bandwidth of a transmitted signal and by doing so, making the signal less prone to a jamming attack.

Jamming deprives the opposition of the use of the radio medium for communications. The only alternative to counter this type of attack is to utilise spread spectrum techniques.

7.3.1 1. Jamming

Jamming (see Figure 7.4) is the term given to the process whereby the jamming station

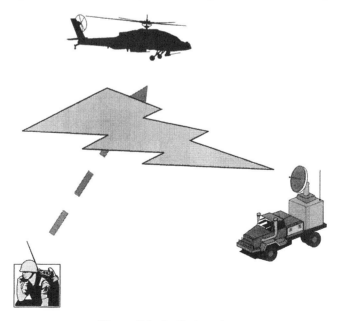

Figure 7.4   Radio jamming

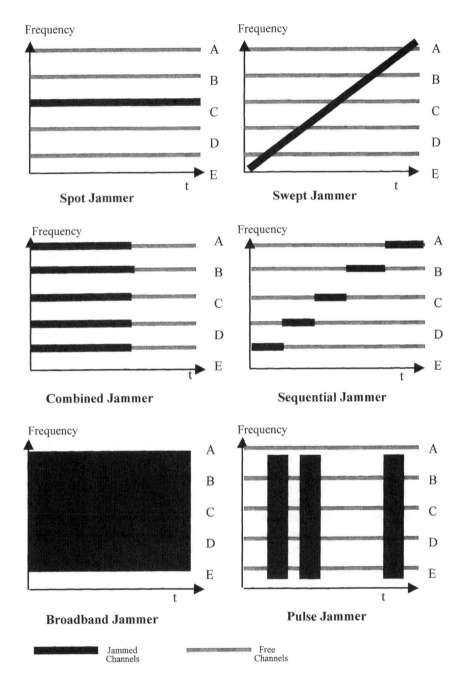

Figure 7.5 Different jamming techniques

occupies the radio frequency band normally used by an opposing force, by transmitting electronic noise and thereby preventing the use of this band for the transmission of, or rather the reception of, transmitted information. Note here, the fact that it is the receiver that is the subject of the jamming and not the transmitter. Anybody listening to commercial radio stations in the HF band, especially during the evenings, is almost certain to experience accidental jamming of their favourite programs by a more powerful transmitter, transmitting at an adjacent frequency. The Voice of America station is one such transmitter that regularly interferes with the lower power transmitters such as those used by the BBC's World Service.

However, in jamming terms this is quite inefficient as most of the desired message will get through anyway, even if accompanied by some annoying noise and gaps in reception. Of course, intentional jamming is a different story and can be much more effective than the example above as it has become a precise and sophisticated technology. Nevertheless jamming is not without its weaknesses and countermeasures. Each of the techniques illustrated in Figure 7.5 have shortcomings as an efficient jammer.

Spot Jamming

The basic principle of spot jamming is establishing the channel being presently used by the target and then transmitting on the same frequency at a power level sufficient to block the receiver station. This would be fine when attacking a fixed frequency station but otherwise the obvious reaction of the jammed link user would be to change to a predetermined and unoccupied channel to continue communications. Thereafter, a cat and mouse game would ensue with the attacker continually searching for the new frequency on which to align the jamming transmitter. The best counter would be for the defender to position his/her channel alongside of that of one of the attacker's channels so that any attempt to jam would now result in some self-jamming on the attacker's part. The latter would be aided by the use of radio channel search and block equipment and a power battle would probably decide the overall winner.

Combined Jamming

A programmed multi channel jammer that transmits on many channels across the bandwidth simultaneously apparently offers a solution. Yet when one considers that in the UHF band, there are around 7000 channels available for selection, then the power required to jam all of these channels makes the ploy untenable.

Swept Jamming and Sequential Jamming

This technique will reduce the power constraints but will only give periodic jamming as the transmitter sweeps through the whole of the bandwidth under attack, transmitting on one channel after another. This would only prove to be disturbing rather than a total block and even jam the attacker's own communications.

Broadband Jamming

Broadband jamming would demand enormous power in order to transmit on some 2000 channels simultaneously. It also produces 'self jamming' as the technique indiscriminately jams all channels, thus rendering one's own radio communications useless.

Pulse Jamming

This technique offers a compromise by periodic jamming transmissions though the bandwidth but would again prove to be inefficient and at best, cause an annoyance rather than a catastrophic loss of communications.

Hence we can see that the jamming of radio channels is not a straight forward matter especially when one considers the degree of jamming that is required to obliterate a channel is perhaps surprisingly high. A spectrum analysis of a jammed channel, of 25 kHz bandwidth will reveal that upwards of 50% of the frequency components must be jammed in order to render the message as unintelligible. Figure 7.5 illustrates the situation graphically.

Nonetheless, despite the difficulties outlined above, it is quite certain that in a time of conflict, the radio spectrum will be seriously compromised with jamming transmissions and in order to maintain a communications superiority, any military force will need to take radical steps to keep its lines of communication open. After all, recent military history has proven that he who rules the electromagnetic spectrum, controls the conflict. Frequency hopping combined with voice message encryption, as discussed in the remainder of this chapter offers the best choice in achieving this goal.

7.3.2 Spread Spectrum Techniques

These include the following options:

- Trunking systems
- Burst transmission
- Direct sequence encoding
- Frequency hopping
- Hybrid systems

7.3.2 1. Trunking Systems

This system overcomes many of the problems brought about by radio jamming. It is a system that uses unused frequency channels, detected by a channel monitoring device for transmission and changes the channel of transmission whenever a jammer engages it. The analogue voice signal is modulated by the output of a stepped voltage controlled oscillator (VCO) which is controlled by the channel detector (see Figure 7.6).

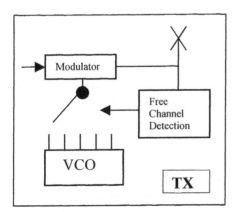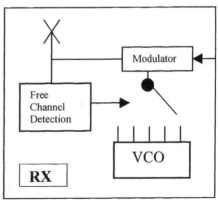

Figure 7.6   Trunking system

7.3.2 2. Burst Transmission

This provides a simple but effective solution (see Figure 7.7). Where greater range is required, burst transmissions, in which a 5-s voice message is compressed by using a buffer memory, with different write/read clocks, into a radio transmission burst of typically 5 ms. This reduces the probability of both interception and direction finding attacks by increasing the rate of data transmitted per second and hence increasing the bandwidth of the transmission.

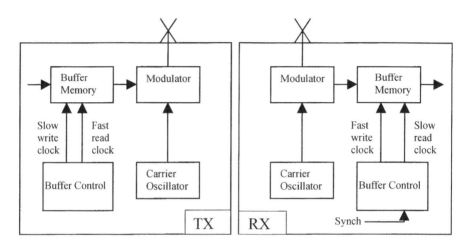

Figure 7.7   Burst transmission

7.3.2 3. Direct Sequence Encoding

The sequence encoding increases the bandwidth of the transmission signal by taking the digitised voice message band of 3 kHz and mixing it with a pseudo-noise digital signal of a higher bit rate. In Figure 7.8 the received message is then extracted from the received composite signal by subtracting the same pseudo-noise sequence at the receiver.

As illustrated in Figure 7.9 the increased bandwidth of the signal makes it more difficult to jam than the original signal. The message spectrum, typically 3 kHz is expanded to that of the higher data rate of the pseudo-noise sequence. It can be seen that the effect of a spot jammer would be minimal, therefore a broadband jammer with all of its overheads would be required to cover the bandwidth of the received signal. DSE exhibits good statistical qualities but suffers from the compromise that for fast synchronisation, short shift registers are required, whereas long shift registers are required for high quality encryption. One further drawback is that it has a process gain in the order of 30 dB and compares unfavourably with that possible from frequency hopping.

7.3.2 4. Frequency Hopping

Frequency hopping (see Figure 7.11) is based upon the derivation of a co-ordinated sequence of hopping frequencies for a 'group of nets of channels' such that collisions (self jamming), do not occur, i.e. the nets are deemed to be 'orthogonal' (see Sections 7.4.4 and 7.4.5). These concepts are further discussed throughout this chapter.

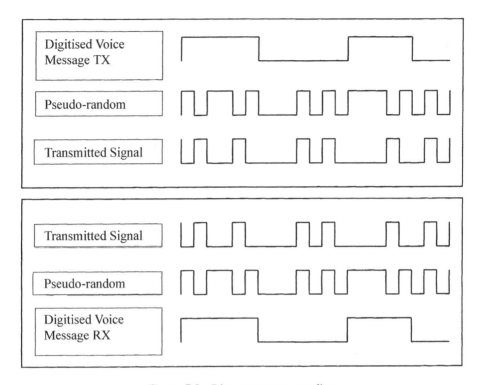

Figure 7.8 Direct sequence encoding

Some clarification of 'groups of nets' is warranted here. Normal radio communications takes place using a channel, of typically, a 25-kHz bandwidth, at a UHF frequency of say 282.900 MHz. If we change the UHF frequency to perhaps 373.025 MHz then we have construed to change the channel and moved the message band of 25 kHz to channel 79. Once again in normal radio communications, a group of channels is known as a 'net', so in frequency hopping where we have a hopping sequence of many channels, we collectively refer to this sequence of 'channel change' as 'nets'. It is more convenient in frequency hopping to consider nets rather than individual channels, as it is the net, which is assigned to the hopping radio link, not channels, as is the case with a fixed radio link. Now it is most likely that a hopping radio network will be able to manage more than one net at a time and hence we refer then to 'groups of nets'. Typically, a good hopping system will be able to handle about 80 nets, possibly arranged in groups of ten, i.e. eight nets in each group. Figure 7.10 illustrates the basic systems approach of a hopping radio unit.

The system illustrated above does not make much sense unless we know what the hopping sequence will be at both the transmitter and the receiver and of course it must be the same in order to retrieve the original message. This means that the pseudo-noise generators must be synchronised at both ends of the link and how better to achieve that than to replace the noise generators with a symmetrical encryption device. This unit would therefore control the hopping sequence and provided the key inputs and the time were the same at both ends of the link, the hopping sequence would be both the same and confidential to the users. Hence

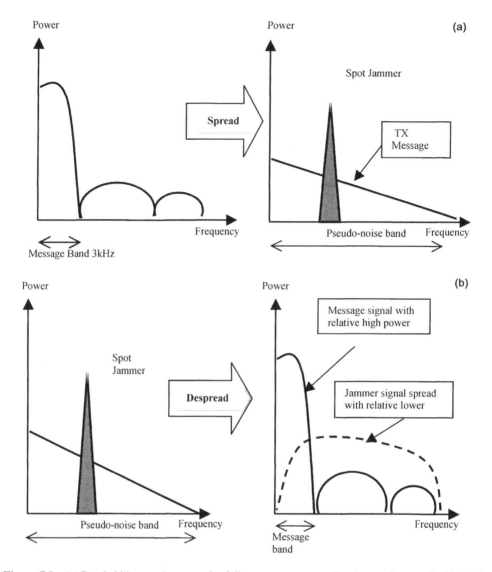

Figure 7.9 (a) Bandwidth spread as a result of direct sequence encoding (transmit-spread). (b) Bandwidth spread as a result of direct sequence encoding (received-despread)

the encryption key data is used to control the hopping frequencies, a highly desirable characteristic for a secure communications transmission.

Of course, now we are 'spreading the spectrum' over the whole radio band that is available and so making jamming largely a redundant pastime. It follows then, that the best channel spread normally available is within the UHF band 225–400 MHz thereby making some 7000 25-kHz channels available, compared with the available VHF band 30–156 MHz which offers far fewer channels.

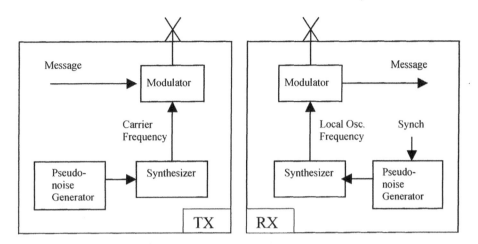

Figure 7.10   The hopping radio system

Note: most military organisations refer to the UHF band as being 225–400 MHz and VHF as 30–156 MHz which contrasts with the otherwise accepted ranges (VHF 30–300 MHz and UHF 300–3000 MHz).

There are even applications where hopping in the HF band is required. The Racal *Panther H* transceiver typically hopping at 10 hops per second over the range 1.5–30 MHz and it follows that there are suitable pieces of hardware available on the market. Even jamming HF transmissions, at this 'low' hopping rate is not without its problems, especially when one considers the power requirement and the frequency agility of the jamming transmitter. The main problem with the HF band is the spectral congestion of the band and devices such as *Panther* use intelligent hopping to overcome this blocking and use single side band (SSB) to maximise spectral efficiency. Whatever the application, it is frequency hopping, which gives us the best tactic in overcoming radio jamming.

A useful quantification of the quality, or effectiveness of a frequency hopping system is known as its *process gain*. The process gain is the $\log_{10}$ of the ratio between the bandwidth available for hopping, against the bandwidth of the message signal. The example drawn in Figure 7.11 gives the following calculation:

$$\text{Process gain} = 10\log_{10}\frac{\text{Bandwidth available for hopping}}{\text{Bandwidth of the message signal}}$$

$$\text{Process gain} = 10\log_{10}\frac{400\text{ MHz} - 225\text{ MHz}}{25\text{ kHz}}$$

Process gain = 36 dB

It is quite clear that operating as a hopper radio in the VHF band and lower will decrease the process gain substantially and conversely using a 12.5-kHz channel bandwidth will increase it. Hence there is a strong argument to hop in the UHF rather than the VHF band.

Adaptive frequency is a useful option, often found in frequency hopping radios. This is a feature which allows the hopping transmitter to sense noisy channels whether they be jammed

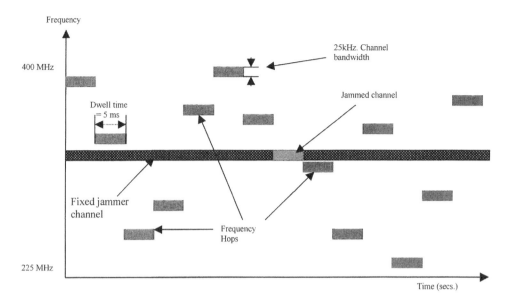

Figure 7.11   The principle of frequency hopping

channels or simply natural noisy ones, and omit them from their predetermined hopping sequence. The effects of jamming or background noise are thus minimised.

7.3.3 COMSEC and TRANSEC

COMSEC from Communications Security, as described in Chapter 2, provides security of the message by encryption, whereas *TRANSEC* protects the transmission medium. A typical basic cipher system is shown in Figure 7.12

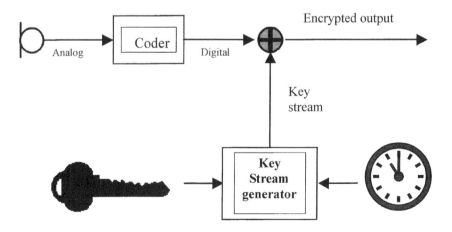

Figure 7.12   Encryption of digitised voice communication

The protection of the radio medium for communications is commonly referred to as *TRANSEC* from TRANSMISSION, or TRANSPORT SECURITY and when combined with *COMSEC* (COMMUNICATIONS SECURITY,) giving message confidentiality and integrity, we find the optimum solution in maintaining our radio communications links. The final situation is summarised in Table 7.1.

Table 7.1 Summary of threats and prevention measures

	Threats			
	Interception	Deception	Direction finding	Jamming
Fixed channel – plain	Possible	Possible	Possible	Possible
Fixed channel + COMSEC	Not possible	Not possible	Possible	Possible
TRANSEC – plain	Possible	Not possible	TRANSEC system dependent	Not possible
TRANSEC + COMSEC	Not possible	Not possible	TRANSEC system dependent	Not possible

7.4 Military Applications

Having established that secure military communications are best served by adopting a combined system of COMSEC and TRANSEC, we need to investigate the requirements of a system user.

7.4.1 Applications Requirements

The application requirements of a combined air, ground and naval force operation demand that for most applications, the frequency range will be in the VHF and UHF bands. The information sources will ideally include voice and data terminals and there will be the essential requirement that hopping radios will need to be compatible with existing fixed frequency equipment. There would often be the need to be compatible with non-encrypted radios. Another important consideration when purchasing any communications equipment is that of the possibility of upgrades in both software and hardware and it would be wise to make provision for this. Radio operations with equipment that does not have these facilities can become quickly redundant in the battle for supremacy between suppliers.

7.4.2 Operational Requirements

The operational requirements dictate that the hopping radios must be 'jam proof'. In the worst case, one might expect to have radio intelligibility with a hopping network being subjected to some 30% jamming. Frequency hopping networks rely heavily upon a strong synchronisation that must be jam proof, exhibit a fast synchronisation time and the posses the ability to synchronise under all conditions such as the 'late entry' into a net activity.

They must allow for 'break-in' whereby certain transmitters, given a degree of priority, are allowed to break into an existing hopping conversation. A squadron leader or base commander would have such a priority programmed in his terminal equipment. The basic operation of an airborne radio must be simple to operate, bearing in mind the complex task of piloting a military aircraft in a hostile environment.

There will be the need for adaptive programmability so that operational parameters can be tailored for specific missions. This then will involve the use of system management, infrastructure and downloading and courier tools.

7.4.3 Security Requirements

The security requirements require the highest possible security, both in the digital encryption within COMSEC and the control of the hopping sequence in TRANSEC by the algorithm. There must be sufficient flexibility within key management to cope with a multitude of scenarios, yet at the same time, the secret key operations should be easy to implement with the burden of the complex security planning being taken over by an efficient and secure management centre. The algorithm must be capable of delivering a long 'hop sequence' and so avoiding any compromising repetition. Most manufacturers will provide protection of the algorithm and keys by various means based on the premise that the operation security should assume that the worst-case situation of loss of keys and algorithm to the opposition, should not jeopardise the mission.

Physical protection of these parameters is especially essential when one considers that there will be inevitable losses of aircraft and other resources within enemy territory. Hence keys and algorithms must be enclosed in an encapsulated, tamperproof module. A simple to operate 'emergency-clear' function is the easiest method of deleting keys when loss of equipment can be seen to be imminent. A further level of support would be that the vital parameters should be protected against unauthorised readout attempts as described in Chapter 1 and a time-out facility whereby secret keys are automatically deleted after the time allocated for the mission.

7.4.4 Anti Jamming Requirements

Anti jamming requirements are that a hopping system must have a high process gain which is achieved by utilising full band hopping over the entire VHF/UHF range, small channel spacing and low data rates. A reasonable hopping rate must be achievable. Now although there is a desire to hop at higher speeds for security reasons, there is the inevitable compromise that the increase in hop rate brings about an increase in the message packaging overheads. Therefore, the radios with the highest hop rates do not necessarily represent the most efficient devices in message transmission, as each hop requires its own packaging and control parameters. The typical present day hopping rates are in the range of 200–500 hops per second. Where greater than 500 hops per second is considered as high speed hopping and below 100 hops per second as low speed hopping. A medium hopping rate is considered the optimal choice in respect of the defining parameters. The limitation of range being due to the requirement of an agile synthesiser, which must be capable of changing frequency at such high speeds (see Figure 7.10).

Frequency management is a critical aspect in frequency hopping systems, as there are the dangers of self-jamming and of co-location effects. Self-jamming can occur when there are several hopping nets in operation at the same time. Without careful frequency management, it is possible that two nets may be using the same frequency channels in their hopping sequence. When such collisions occur, self-jamming will be experienced. The solution in frequency management is to arrange the hopping operation by the use of *orthogonal nets* where individual hopping nets are co-ordinated so as to avoid self-jamming. This is not an easy task without a suitable management tool or centre.

7.4.5 Co-location

Co-location interference occurs when the operation of a number of transmitters and receivers is simultaneous and within close proximity to each other as is typically the case at HQs, command centres, air bases and on naval ships. The problem is that self-jamming occurs due to interferences such as *intermodulation, spurious transmission* and *blocking*.

The problem becomes critical when a transmitter and a receiver operate from the same site and use the same hopping band. A reduction of interference levels in the order of 50 dB can be realised by separating the antenna by 20–50 m. Where this is not possible due to other operational constraints, then an alternative ploy might be to use fixed frequency transmission for the short links, as they are less likely to be vulnerable to jamming. A second alternative would be to rotate antennas so that one is using vertical polarisation whilst the other uses horizontal polarisation. In this way a signal to noise improvement of > 10 dB can be achieved.

The effects of co-location can also be reduced by the tactic of hopping within sub-bands as implied in Figure 7.25. If neighbouring transmitters are arranged so that one transmits in the higher band with the other transmitting in the lower band, then shared site stations can benefit enormously.

To avoid virtually all co-location interference between neighbouring stations, a noise-to-signal ratio of > 150 dB must be achieved, but this can only be accomplished by a separation of > 2 km.

So the solutions to the anti-jamming requirements are to adopt:

- Co-ordination of frequencies by developing *orthogonal* nets.
- Meticulous frequency management.
- The geographic separation of transmitters and receivers.
- A medium hopping rate.
- Adaptive power transmission.

All of the above requirements have one further criterion to meet, in that they must be satisfied within the budget available!

7.4.6 Air Defence Scenario

In the air defence scenario (see Figure 7.13), the maximum range of about 350 km is to be expected with typically, a maximum number of 80 communication links.

The aircraft mission is controlled from the air operating centre (AOC) with a provision to hand over communications to supporting AOCs should the aircraft move out of range of the original. Alternatively the original AOC can maintain command by means of a communications backbone and for local control a mobile station can be utilised.

With multiple strike forces operating at the same time, each force, or squadron, in the case

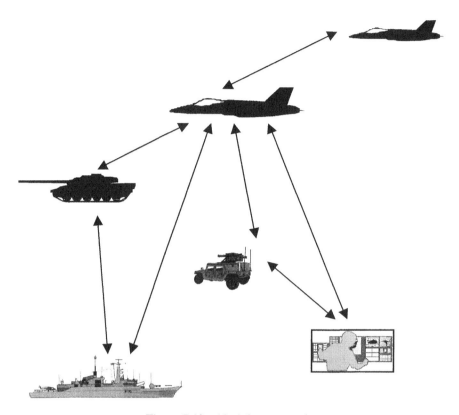

Figure 7.13   Air defence scenario

of an airborne force, is likely to be operating on a different radio net, i.e. hopping sequence, for different parts of the mission and a common net for pan-force communications. All of these mission features and tactics must be carefully planned and programmed into each participating radio unit. So that the only action that needs to be taken by pilots and other operators, is to merely select a 'channel' on the radio. Each channel will have been preloaded before the mission, with operating parameters, key data for the COMSEC channels and key driven hopping sequences for the TRANSEC nets. It should be remembered that in the latter case, a TRANSEC net consists of a number of channels through which the radio will hop. Hence each sequence of the mission will probably require a different TRANSEC net, or fixed frequency COMSEC channel to be selected. Figure 7.14 illustrates the basic net/channel selection situation available to a pilot.

Mode Switch Position

1. Plain transmission channel zz
2. COMSEC channel xx
3. COMSEC channel yy
4. TRANSEC Net tt
5. TRANSEC Net rr
6. TRANSEC Net ss

Figure 7.14   Typical airborne radio control module

As we shall see later, the whole network can be thought of as being controlled by the secret keys as each mode switch selection, apart from the plain channel, is actually selecting the key allocated to that function. Moreover, in the TRANSEC cases, the key determines the net of hopping sequences. Hence, the network management is actually cryptographically structured, once the available frequencies have been identified.

7.4.7 Close Air Support Scenario

In the close air support scenario, the opposing forces will, almost certainly jam communications between the forward air control (FAC) and aircraft over the forward edge of the battle area (FEBA) or enemy territory. Therefore frequency hopping would be an essential requirement. The application introduces the possibility that ground troops, supporting the air forces may only have fixed frequency radios at their disposal. In this situation, for the fixed frequency units to be able to communicate with the hopping aircraft radios, the latter must be equipped with a *hailing* facility. Hailing is a technique whereby the fixed frequency radio transmits on one of the predetermined hop channels programmed into the hopping radios. Typically when the aircraft radio hops onto that channel, the pilot will be warned by an audio tone, that a fixed frequency station is hailing him. He will then have the option of either ignoring that call or switching to a pre-assigned fixed frequency channel, on which he can communicate in COMSEC mode or in plain. It follows then, that the hailing channels must be included in the hopping sequence planning. The technical details of *hailing and break-in* as well as other types of hops are discussed further in Section 7.5.

Figures 7.15 and 7.16 merely illustrate some of the possible communication links that might be planned into a mission profile, but they give little indication of the complexity of a real communications network architecture and management. There are two points of view, which deserve much consideration when designing a hopping radio net, i.e. the radio physics and deployment and that of the security. It is the latter, which is of greater interest as far as the specification of this book is concerned. Though it should be remembered, that the two sides to

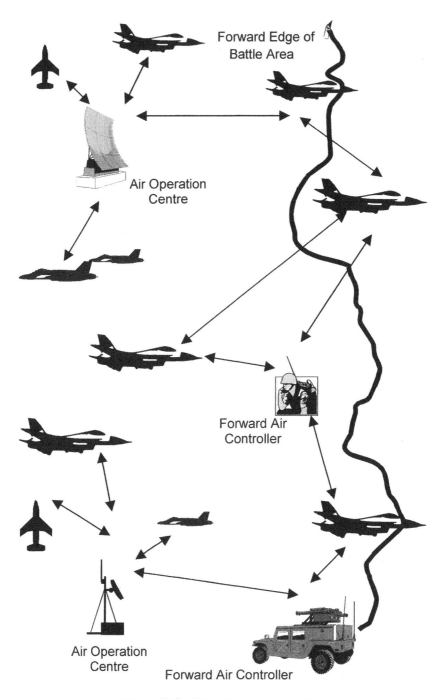

Figure 7.15 Close air support scenario

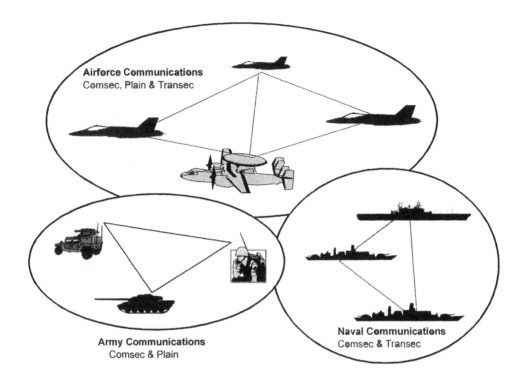

Figure 7.16 Multiforce operations

the story, are not separate entities. Indeed, the network manager must be aware that each plays an integral part in designing workable and efficient network architectures.

7.5 Network Architecture and Management

If we consider the Tri-Force model in Figure 7.16 in some detail, we would find perhaps, the following application requirements for a joint operation and that network management can be thought of, in terms of secret key assignment.

The airforce has three possibilities as far as communications are concerned. Each aircraft is able to transmit and receive in Plain, COMSEC and TRANSEC modes. The ground forces, for the sake of this exercise, are deemed to be only capable of using Plain and COMSEC modes, having largely, only fixed frequency radios available for this mission, with the exception of a few hopping radios in the FEBA. Naturally TRANSEC facilities would be the norm. Meanwhile the navy is seen to operate in either in Plain, COMSEC or TRANSEC modes.

A further constraint is that each radio unit has a limit of eight secret keys with which to operate on various channels or hopping nets, during the course of the operation. These do not need to be the same in all units, just as long as there are sufficient keys for inter-force use. Secret keys used in inter-force communications must be common to those parties who wish to communicate.

Table 7.2 Typical channel/secret key assignments

Secret Key	Air Force			Navy			Army		
	Name	Mode TRANS (T) COMS (C)	Net / Channel	Name	Mode TRANS (T) COMSEC (C)	Net / Channel	Name	Mode TRANS (T) COMS (C)	Net / Channel
SK 1	Squad. 111 All air	T C	Net 1 Ch. 1-5	Pacific Flt.	T	Net 1a	Comp.1	C	Ch. 20-25
SK 2	Squad 202	T	Net 2	Atlantic Flt.	T	Net 2a	Comp.2	C	Ch. 25-30
SK 3	Tower	T	Net 3	Port	T	Net 3a	Battal'n	C	Ch. 30-35
SK 4	Base HQ	T	Net 4	Base HQ	T	Net 4a	Brig.HQ	C	Ch. 35-40
SK 5	Air/ Ground	T	Net5	All Navy	T	Net 5a	FAC Ground/ air	T	Net 5
SK 6	Air/ Ground	C	Ch. 6-10	Navy/ Ground	C	Ch. 6-10	Ground/ Air /Navy	C	Ch. 6-10
SK 7	Air Navy	C	Ch11 -14	Air/ Navy	C	Ch11-14	Ground/ Navy	C	Ch. 41-43
SK 8	Air Navy	T	Net 6	Air/ Navy	T	Net 6			

☐ = Common SK's

It is inevitable that each force will have its own hierarchy of secret key (SK) management using SKa for the airforce, SKn for the navy and SKg for the ground forces, such as those suggested elsewhere in this book (Table 7.2).

Note here, the distinction between an SK controlling the hopping sequence of a TRANSEC net and an SK encrypting a COMSEC channel and also that the SKs which are not common to more than one force, can have different values. It is also worth noting that the radio communications will have up to, typically 80 hopping nets and many more fixed frequency channels available. However, as it will be seen, the 80 hopping nets will almost certainly be arranged in groups, as in Table 7.4. There is no reason why an SK cannot be used for more than one purpose as inferred above, to provide an alternative secure communication channel and mode. It would be wise to provide both TRANSEC and COMSEC options for an application to give cover for some possible unforeseen problem arising with the frequency hopping system. Of course, there are many possible permutations to work with, in tailoring the key management to match mission profiles.

Undoubtedly both the ground and naval forces will have a similar SK hierarchical structure to that of the airforce. Therefore, as far as the joint operation communications between forces are concerned, the overall picture would now look like that in Figure 7.17.

7.5.1 Mission Procedures

Assuming that all transceivers are loaded with their specific SK data, the first step for the airforce units in their pre-flight communications set up, would be to perform a net entry. A net

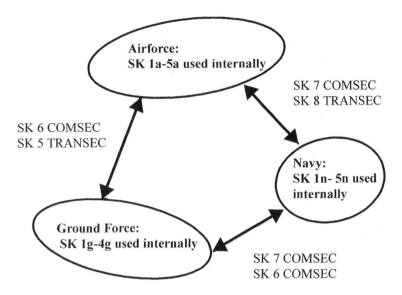

Figure 7.17   Secret key assignment in the Tri-Force scenario

entry is usually carried out at the airbase, in a jamming free environment and allows all hopping radio units entering the net, to be able to synchronise their hop clocks by taking their time setting from a unit declared as a time master. This is a role usually played by the airbase tower or the squadron leader. Any pilot or controller failing to synchronise at this point will be unable to communicate later in the mission, using the hopping system. The same procedure must be carried out by the navy's hopping radios and also steps must be taken to ensure that joint airforce/navy hopping radios are synchronised too.

For the COMSEC radios, critical time synchronisation is not a cause for concern unless they are utilising a *time authentication* function. In which case, accurate time settings *are* critical for COMSEC use and communications between these units should be checked for successful operation before the mission commences. A further test to be carried out is that of checking the *hailing operation* between fixed frequency and hopping radios. In the exercise above, the ground force radios will need to switch to a pre-determined plain channel, which would be exclusively used for hailing between a ground station and an aircraft or naval vessel. On hearing the audio hailing alarm, the pilot must switch from the hopping net, to the hailing frequency and then to an agreed COMSEC channel to enter into ciphered communication with the fixed frequency station.

All of the pre-ops procedures must be carried out prior to the mission and in a radio friendly environment, distant from the listening ears jamming transmitters of the enemy.

7.6 Characteristics of Frequency Hopping Networks

7.6.1 COMSEC

The shaded components of Figure 7.18 show the COMSEC components of an integrated frequency hopping radio controller. The operation of a COMSEC processor is described in

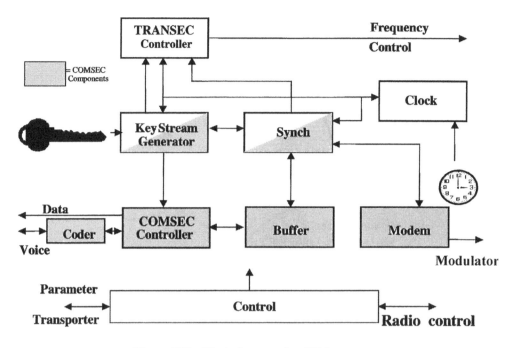

Figure 7.18 Block diagram of an EPM processor

detail in Chapter 2 but here it can be seen to share the Key Stream Generator and the Synch Function, with the frequency hopping processor, even though their functions are somewhat different. With the COMSEC operation, the Key Stream is used directly to encrypt the voice signal whereas in the TRANSEC operation, the Keystream is used as the controlling function for the hop sequence. The synch operations are also different. In COMSEC, the synch function is used to maintain the transmitter and remote receiver keystreams in step. However, in TRANSEC, the purpose of the synch is to dictate the position of the synch hop and the data that the hop contains. What is common between the two entities, however, is that they are both dependent on time and keys. Any variation in these parameters in different radio units, will rule out either COMSEC or TRANSEC operations, depending upon other parameter options. A second common feature is the *control function*. This block represents the uploading and downloading operations of key and parameter data between the EPM processor and the key management centre and, of course, the transceivers themselves. This subject will be discussed in greater detail in Section 7.7.

7.6.2 TRANSEC

The block diagram of Figure 7.19 expands the basic TRANSEC functions shown in the previous diagram. Now we can see that there are four essential parameter inputs into the TRANSEC generator.

- The SKs, typically eight in number
- The accurate time

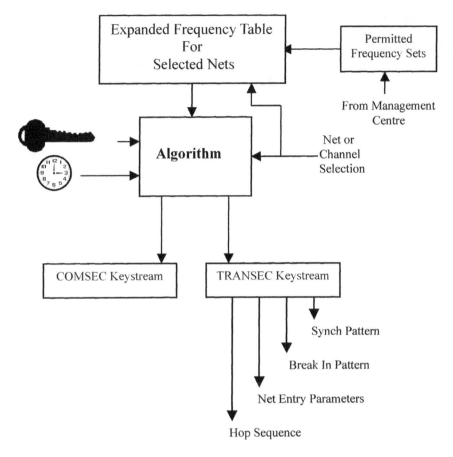

Figure 7.19   TRANSEC generator

- The net or channel selection from the radio operator
- The permitted frequency sets

These are the parameters, which the algorithm needs to generate the controlling outputs:

- The synch pattern
- The break-in pattern
- Net entry parameters
- The hop sequence
- The COMSEC keystream

'Break-in' and 'net entry' are further discussed in Section 7.6.2.2. See Section 7.5.1 and Figure 7.23.

7.6.2.1 TRANSEC Keystream

Synchronisation

The synchronisation procedure is essential to the frequency hopping system and the development of a good synch system is certainly an engineering challenge and so the actual process is often the closely guarded trade secret of the manufacturer. Therefore, the procurer must be content in the knowledge that the synch sequence, as indeed the other hop types, are defined by the system time, the secret keys and the algorithm. If the synch sequence were in any way predictable or on a fixed frequency, then it would be vulnerable and a certain target for the jammer. Once synch is jammed, the whole hopping system would be inoperable. The question is, how can the genuine receiver be certain about the timing and the frequency of the next synch hop? The solution relies upon the common knowledge, within the friendly network, of the system clock and the secret keys. The threat and solution are illustrated in Figs. 20 and 21.

It is most desirable that the synch time is as short as possible in order to keep the transmission overheads to a minimum and yet robust enough to overcome noisy channels or jamming attacks. Similarly, synch must occupy only a minimum channel capacity. It must also provide for *late entry* capability when a transceiver wishes to join an on-going conversation between other stations and must be resistant to spoofing or replay attack as described in Chapter 1 as well as being fully automatic and transparent to the user.

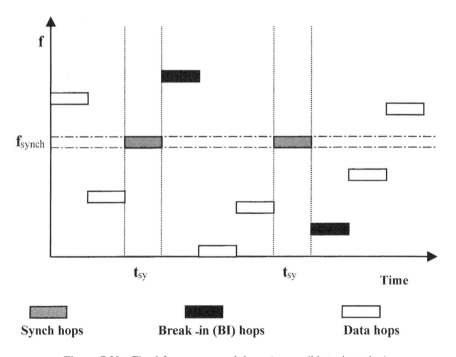

Figure 7.20   Fixed frequency synch hops (susceptible to jamming)

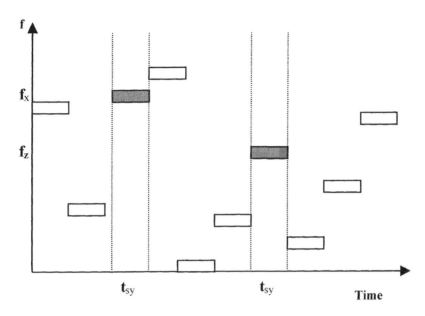

Figure 7.21   Hopping synch controlled by time and SKs (jam free)

7.6.2.2 Synch Hops

The 'search for synch' is illustrated in detail in Figure 7.22.

The transmitted synch hop contains a pattern of data that is specific to the friendly net so that false synch with other nets operating in the region is avoided. Non-transmitting stations are usually in a search mode and in this state and these receiving units, monitor the expected synch frequency for the synch hops. When a synch hop is detected, the receiving station adjusts what is known as its *foreground clock* (dt1 and dt2 in Figure 7.22), to that of the transmitting station. Hence the following payload, along with various 'control' hops, is received in a synchronised manner. Once the communication has ceased, the receiving station falls back to its original timing as per its own *foreground clock*. There is a limit, however, as to the ability of receiving, or searching stations to adjust to a transmitter's timing. Should unit foreground clocks be outside the radio's specification of typically 65 ms (depending upon the hopping rate and the system type), then synch will fail. The *late entry* facility works on the principle that should a station miss the initial part of a communication, then it will be able to join it at a later time as long as its own *foreground clock* time is within the specified limit.

The entire operational timing is dependent on the system's *background clock*. This clock must be set and loaded into all stations participating in the hopping nets before a mission commences and it is crucial that all the hopping radios should have a background time difference of typically, less than 2 min. There are numerous methods of achieving this accuracy, but clocks do drift with time, particularly those installed on aircraft and if a unit drifts beyond the specified limit, then it will not be able to synch with the rest of the net. Hence the use of a time master station, probably that of a base HQ or in the case of squadron of aircraft, the squadron leader, to control the net time. It would be a standard procedure during the mission immediate preparations, when all the hop and key parameters have been

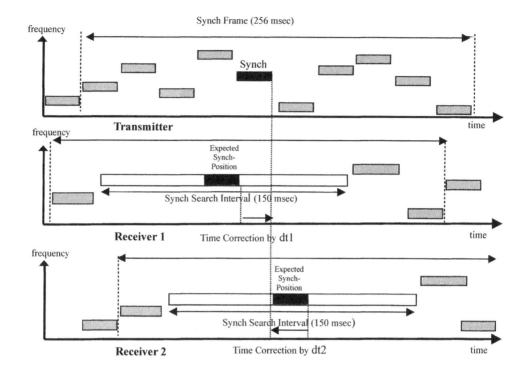

Figure 7.22   The search for synch

loaded into the aircraft radios, for the mission pilots to perform a *net entry*. In carrying out this procedure, all units will synchronise their background clocks with the single radio that has been designated as the time master. The management of time needs careful consideration and is further discussed in Section 7.7.6.

The *break in* BI hop, which is similar in principle, but different to the *late entry hop*, carries information, which enables a station wishing to break into an existing communication, to do just that. The break in station, which is initially a receiving station, transmits the BI hop at the correct moment in the synch frame and in subsequent frames over a period of several seconds to ensure its reception even under fading or jamming environments. On receiving the BI pattern, all units, including the original transmitter will synchronise their foreground clocks with that of the BI station and will then switch to search mode, enabling the BI station to take over the net as the new transmitter. Units in the net initiate a guard period for several seconds after a BI to prevent the reception of the other BI hops interfering with the communication link.

A *net entry* (see Figure 7.23,) is carried out by any slave station in the search mode.

The slave sends a time request (TRQ_1) to the master, which also must be in the search mode. The master then reacts by sending a time response (TRP_1) to the originating slave and any others who are in the search state. The response is repeated on the next hop frequency as TRP_2 and this results in the receiving slaves being synchronised with each other, providing that their background clocks are within about $\pm$ 2.5 min depending upon the system's specification.

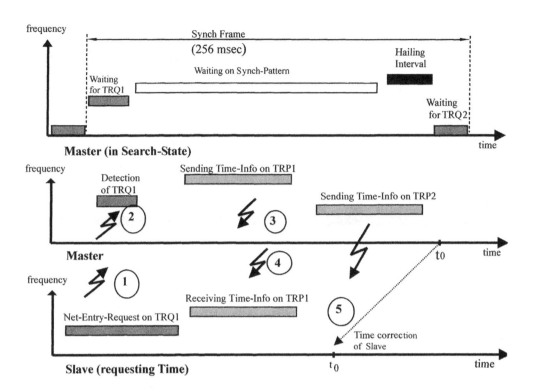

Figure 7.23   Net entry, time requests and responses

Hailing, as mentioned earlier, is the method by which a fixed frequency radio can communicate with a hopping radio.

Prior to a mission, the frequency assignments for the hopping radios, the frequency of any fixed frequency station must be included in this assignment if access to a hopping radio, by a fixed-frequency station is required.

Hailing hops, as in Figure 7.24, are performed by a hopping station in the search mode. As per the programming, the hopping station will periodically hop onto the fixed frequency channel and 'listen' for any signal on that frequency. On the reception of a plain fixed frequency signal on the hailing frequency, an audible alarm can be programmed to alert operators that a fixed frequency unit is 'hailing' them. When this occurs, the operator of the hopping station must choose to switch to plain mode to communicate with the hailing station, if he wishes converse. Having done so, then it would be prudent for both parties to switch to a common COMSEC channel to carry out the ensuing conversation. In order to avoid false detection of hailing, a squelch tone must be present in two out of three consecutive hailing hops. If it is inconvenient for the operator of a hopping station, e.g. a pilot, to switch communication modes, then he would be free to ignore the hailing call if he so wished.

7.6.2.3 Frequency Assignment

With no net co-ordination of a hopping system, collisions of channel use will occur as

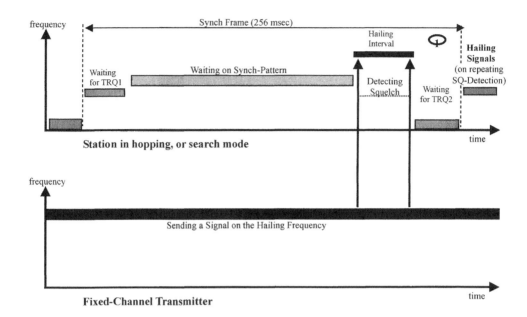

Figure 7.24   The principle of hailing

overlapping nets hop onto the same frequencies simultaneously. This will result in *self-jamming* and the resulting interference could cause loss of communication. One tactic, that could be used to overcome this problem, called *static co-ordination,* (see Figure 7.25) divides the available bandwidth into sectors so that one net would only be allowed to use frequencies within a certain sector. This certainly overcomes self-jamming, but at the cost of drastically reducing the process gain and hence leaving the net more vulnerable to aggressive jamming attacks.

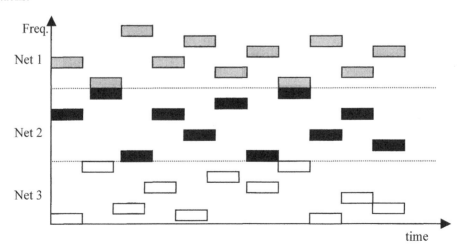

Figure 7.25   Static co-ordination

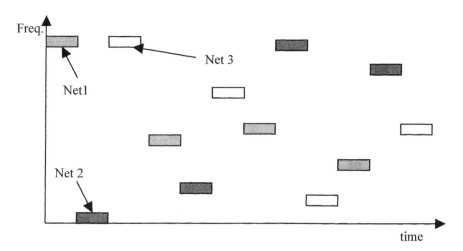

Freq.

Net 3

Net1

Net 2

time

Figure 7.26   Dynamic co-ordination

A better solution would be to carry out what is known as *dynamic co-ordination* (see Figure 7.26). In this case, the frequency planner draws up a table of channels covering the whole bandwidth in use. From Table 7.3, which represents just a fraction of the 7000 channels available, he would need to eliminate all those channels in use by other networks or carrying functions such as guard channels for instrument landing systems and emergency channels, e.g. 243 MHz. The latter also requires 25 kHz guard channels either side of the designated frequency. Once he has done this, the frequency planner can step through the remaining channels and allot them to a frequency set for use by a particular net. Planning should take particular care to include the hailing frequencies in his net. F.Set 0 and F.Set 1 are seen to include hailing frequencies, whilst those of F.Set 2 and F.Set 3 are without, thereby implying that these latter sets will only be used in hopping nets.

From Table 7.3, it is interesting to note that F.Set 4 includes channels already included in other sets. On first consideration, this would seem to be a contradiction of the previous discussion. However, this selection is permissible providing that the geographic distance and therefore radio range, between the nets using those frequency sets is sufficient to rule out the possibility of self jamming. Hence, there is a degree of frequency/channel reuse, just as takes place in the cell planning in GSM telephone systems.

7.6.2.4 Key/Frequency and Channel Assignment

Frequency hopping systems vary according to the manufacturer and in particular with the processor power available, so the arrangements of frequencies and channels into nets and groups may differ. However, it is likely that the arrangement discussed here is a typical example in that the processor can control 80 nets, remembering that in frequency hopping terms, a net is a hopping channel. These nets are arranged in eight groups of ten with each net group being assigned a frequency set such as those designed in Table 7.3. The result is a table of groups as shown in Table 7.4. So the pilot of an aircraft will only need to dial, or punch in the group number digit by means of a keyboard and the radio will automatically select the

Table 7.3 Selection of channels for frequency sets

UHF-Band MHz	Frequency Sets							
	F. Set 0	F. Set 1	F. Set 2	F. Set 3	F. Set 4	F. Set 5	F. Set 6	F. Set 7
225.000	▨				‖‖‖			
225.025		▨						
225.050	Used	Used	Used	Used	Used	Used	Used	Used
225.075			▨					
225.100				▨				
	▨				‖‖‖			
228.775		▨						
228.800			▨		‖‖‖			
228.825	Hailing	▨						
228.850								
				▨	‖‖‖			
243.000	Blocked	Blocked	Blocked	Blocked	Blocked	Blocked	Blocked	Blocked
		▨						
398.425		▨						
398.450			▨					
398.475				▨				
	▨				‖‖‖			
			▨					
400.000				▨				

Table 7.4 Table of groups

Index/group	0	1	2	3	4	5	6	7	8	9
0	00	01	02	03	04	05	06	07	08	09
1	10	11	12	13	14	15	16	17	18	19
2	20	21	22	23	24	25	26	27	28	29
3	30	31	32	33	34	35	36	37	38	39
4	40	41	42	43	44	45	46	47	48	49
5	50	51	52	53	54	55	56	57	58	59
6	60	61	62	63	64	65	66	67	68	69
7	70	71	72	73	74	75	76	77	78	79

appropriate settings of key and frequency set. In COMSEC mode, the pilot will need to punch in the two digits representing the appropriate channel required and the assigned secret key is automatically selected for that channel.

Table 7.5   Group/channel key assignment

TRANSEC

Net group	Secret key	Frequency set
0	1	1
1	5	4
2	5	5
3	2	2
4	7	3
5	5	8
6	8	7
7	3	6

Having established the net group numbers, there remains the task of allocating the secret keys to those nets, in TRANSEC (see Table 7.5) and to the channels in COMSEC (see Table 7.6).

Table 7.6   Channel/net-SK assignments

COMSEC

Channel net	00	01	02	..	..	77	78	79
Secret key number	4	6	8	..	..	5	1	0

7.7 Key/Data Management and Tools

As suggested in Chapter 2, cryptographic aspects such as algorithms and key lengths are not the only factors to debate when assessing the qualities of communications security system. For such a complex system as a frequency-hopping network, with multi-mission facilities, a high quality management centre is essential to manage the keys and parameters of all of the network components. Without efficient management tools, the best encryption processes available for frequency hopping, can be rendered at best, inefficient, if not totally useless. A sensible approach to the use of data sets, is to assemble two complete sets, which may only differ in secret key data, but the policy has the advantage that when two or more data sets can be distributed amongst network components at the same time, continuity of operation is enhanced when a quick change of data set is called for. In this way, a sudden change of mission is more easily catered for.

The management of a frequency hopping system can be broken down into four parts.

- Secret key data
- Frequency data
- Set-up data
- General operating parameters

7.7.1 Algorithm Data

As indicated in Figure 7.27 the algorithm requires input data for both the COMSEC and TRANSEC keystreams.

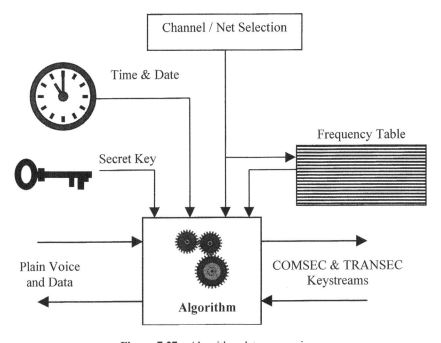

Figure 7.27 Algorithm data processing

The COMSEC data input to the algorithm includes the SK, which is one of eight SKs that are used for the COMSEC encryption process, as selected by the choice of radio channel. A management centre will need to have the ability to generate keys from a true random number generator source and a facility to generate and store passwords used for accessing station functions. *Date* and *time* are included for the initialisation vector of the key generators and also for 'replay mode.' The *channel/net* selection that designates which SK is used for the encryption of a specific channel.

The TRANSEC components are made up of the *channel/net* selection, which in this case, is used by the frequency table to derive the frequency and hop data associated with the selected net. The frequency table provides the frequency sets, the orthogonality and the various operating parameters used such as hop rate and dwell/time ratios, and finally the *date* and *time,* which are used in TRANSEC, for the synch frame structure, which was described in Section 7.5.

7.7.2 Frequency Data

The frequency data comprise the required operating band, usually VHF or UHF, those frequencies in the frequency range to be used by the hopping system and their assignment to the channels or nets, as well for special functions such as hailing. For COMSEC operation, the allocation data for frequency to channel assignment is required. Other frequency parameters include the channel spacing, frequency offset of 12.5 kHz, if required, the blocked channels, the hop rate and dwell/time ratio. There is also the possibility of a TRANSEC control function being included, to ensure that the dispersion of hops throughout the whole of the available bandwidth. The function is not described in any detail in this book and it suffices to relate, that a frequency 'spread' could be applied to the hop set to remove the possibility that hops may be concentrated within a narrow band, which would leave a hopping system more vulnerable to jamming.

7.7.3 Pre-set Data

The pre-set data largely determine the role that each radio station is allowed to play in the ensuing mission, according to the mission manager's requirements. Such data would include, the operating mode of the station, i.e. COMSEC, COMSEC/TRANSEC with or without hailing, voice or data, or voice and data, channel or net number and the hailing frequency. In addition, the type of modulation (AM or FM), station role (either slave or master,) break-in capability, cipher or plain override and the option for a plain marker would be added.

7.7.4 Configuration Parameters

The installation set-up of a radio requires a number of parameters to determine how it will operate, whether as a transceiver, as a split site station or as a relay station with/without monitoring facilities. Stations will also need radio specific data such as transmitter power, audio gain and should include a *time out* for the cryptographic data. This parameter controls the length of time, e.g. mission duration, or after the power has been switched off, for which secret keys, etc. are stored in a unit, before being removed.

A typical management centre will bring all of these inputs together so that they can be co-ordinated, stored and packaged securely for distribution throughout the network (Figure 7.28). This may be carried out by courier, or by *down line loading* over secure radio or telephone.

7.7.5 Key Distribution

The *data courier device* (DCD) is a storage and distribution device, which can upload complete data sets from the management centre and subsequently download the data sets

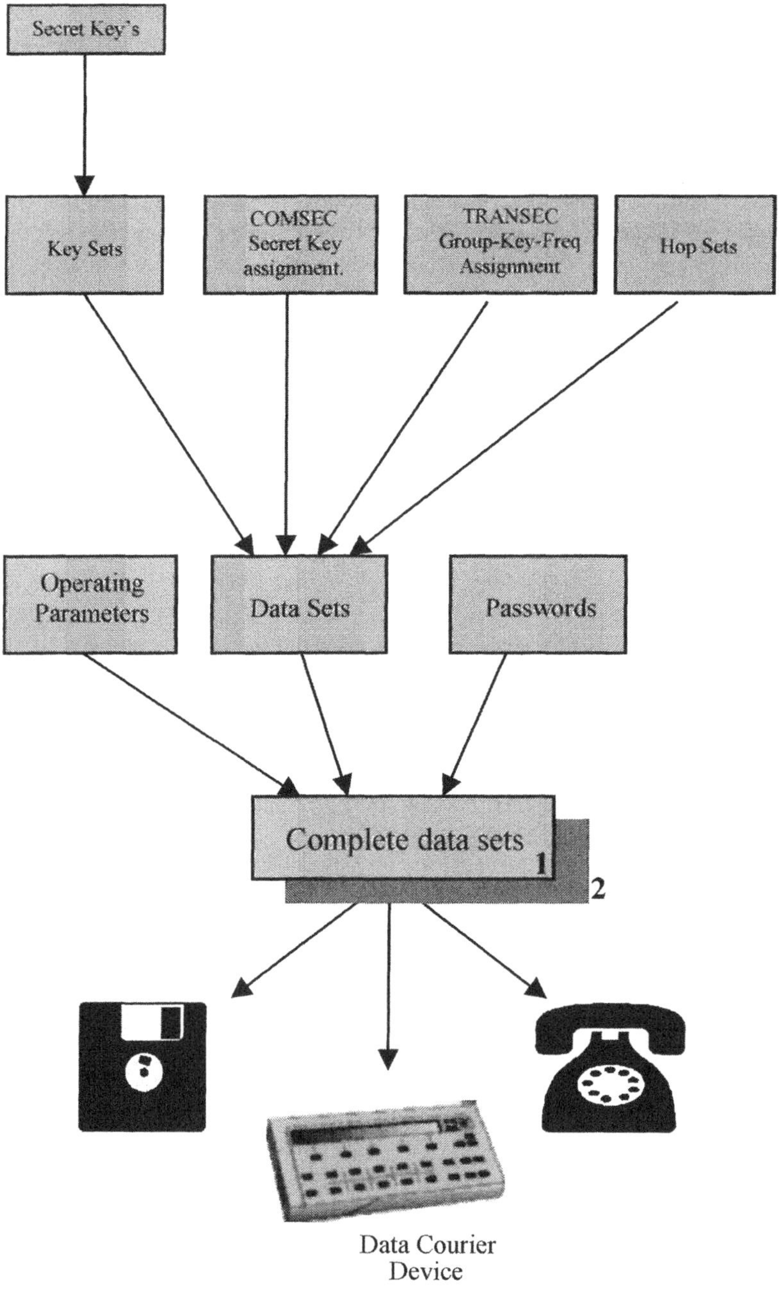

Figure 7.28   Functions of a system management centre

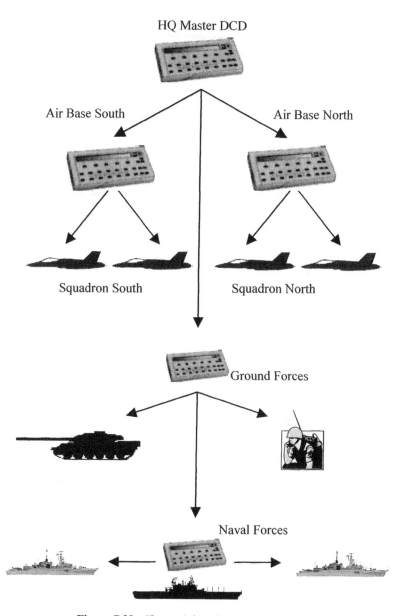

Figure 7.29 Key and data distribution by DCDs

to individual radio units or management centres or indeed clone to other DCDs to allow rapid data distribution to multiple and widespread centres. Providing sufficient DCDs are available, perhaps the most efficient use of these devices would be to employ them as base resident machines, with rotation around a management centre to facilitate convenient data set updates. One suggestion is shown in Figure 7.29.

Access to its operation must be protected by a password hierarchy, which may allow some editing of the data sets stored in it. It must also be further protected by a read out limit, emergency clear and time out functions in case of its loss or capture. A DCD should also provide feedback data on its operational history to the system manager, so that he can monitor the device's efficiency and secure use.

7.7.6 The Time Problem

As we have seen, accurate time management is vital to frequency hopping systems and it becomes more difficult, as networks get larger, particularly when they are dispersed over great distances as might be expected in a multi-force operation. An apparent solution is to use the GPS time by linking GPS receivers to radio units as used in the 'Have Quick' frequency hopping system, but in practice this is not as easy as it might seem. There are several problems which can result in stations, particularly airborne units being left out of the net due to a failure to time synchronise and include radio/GPS incompatibility, faulty or inaccurate GPS data due to lack of satellite data and of course user error.

Manual time inputs, whilst seemingly tedious and dated, are still a reliable and flexible option. By the nature of their construction, frequency hopping ground stations are not so restricted in the size and facilities as are airborne radios. This means that the airborne clocks are more vulnerable to time drift than the more 'complete' ground units. Hence, it makes sense to rely more on a ground station as a master than on an airborne unit although for the sake of an airborne mission, the squadron leader's aircraft should assume the time master station when the flight is out of VHF/UHF range with the base.

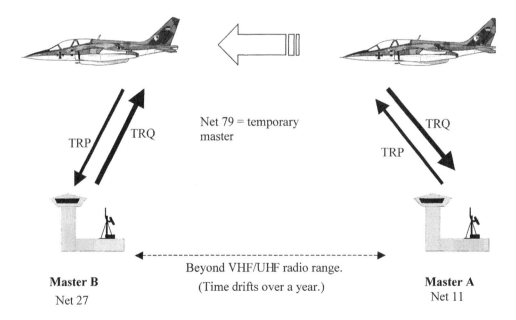

Net 79 = temporary master

TRP   TRQ

TRQ   TRP

Beyond VHF/UHF radio range.
(Time drifts over a year.)

Master B   **Master A**
Net 27   Net 11

Figure 7.30   Spreading the system time by airborne master

The situation becomes more complicated when squadrons from two or more bases are involved in a mission. The same can be said for dispersed naval and ground forces. Where the airbases are within radio range, then there is little problem in arranging for one to act as time master so that it can transmit its own time to its slave partner when it performs a net entry along with the normal TRQ/TRP handshake. The slave station at the second base then reverts to its normal time master status and thereafter controls the time of its own aircraft. It follows then, that all of the aircraft participating in the mission, will have accurate clock so that synchronisation between the squadrons can be relied upon.

When bases are remote from each other, however, the situation becomes a bit trickier and it is necessary to spread the system time by the use of an airborne master. This is illustrated in Figure 7.30. Master A is assumed to be the controlling station and Master B, a remote airbase, out of VHF/UHF range of Master A, but required to have the same timebase, remembering from previous arguments, that should their aircraft have time differences greater than, say, 2 min, then they will not be able to synchronise their hopping radios. The procedure for a *time master flyby* is as follows:

1. Master A is normally using net 11, but Master A and the aircraft switch to net 79 for the calibration flight set-up.
2. The aircraft, having carried out a net entry with Master A, now holds the exact system time.
3. Master A reverts to net 11 for its communications with other units.
4. The aircraft switches from its 'normal' slave state to become a temporary master and flies to within range of Master B.
5. Master B switches to net 79 and carries out a net entry procedure, as a slave, with the airborne time master.
6. On completion of the net entry, Master B has assumed the time of the aircraft and now switches back to its normal net, i.e. 11 and its mode of operation to master. Subsequently, Master B will be able to dispatch its new time to its own squadrons, confident that they will have the same time as those squadrons from Master A base.
7. Meanwhile the airborne time master returns to its base, switches back to slave mode and rejoins its communicating partners on net 11.

7.8 Hardware Components

Figure 7.31 shows Panther V frequency hopping radios configured as a ground base station.

7.8.1 Airborne Transceiver

The control module in the case of fighter aircraft is of course located in the cockpit with the main module installed in the fuselage, with access panels available through which mission data can be uploaded by means of a DCD.

The typical data specification is described as follows:

Figure 7.31   Panther V hopping transceivers configured as a base station (courtesy of Thales Defence Ltd.)

Environmental data

Airborne station:	According to MIL-E-5400
Operating temperature	− 50/ + 70°C
Storage temperature	− 50/ + 90°C
Shock temperature	minus;40/ + 70°
Low pressure	50 mbar@15240 m
Humidity	+ 60°C/95% RF
Vibration	5 g
Shock	10 g
Drop	1.2 m

Technical data

Frequency range	VHF 30/174 MHz (split bands)
	UHF 225/400 MHz
Modulation	AM/FM/FSK
Channel spacing	25 kHz
Hopping rate	150/500 hops per second
Dwell time	4/6 ms
Frequency	
Switch-over time	0.5 msNumber of hopping
Frequencies	7000 max in UHF bandTX/RX switchover
Time	2 ms
EMC	MIL-STD 461 B Class A3
TEMPEST	Propriety standard
ECCM mode	Stanag 4246
Built-in test (BITE)	Mil-STD-2084

Cryptographic data
Standard or proprietary algorithm
Number of secret keys   2 sets $\times$ 8–16 keys
Secret key diversity $10°32$ each
Key stream period > 9000 years
Initialisation vector diversity 2^{24}
Time authentication

8

Link and Bulk Encryption

This chapter looks at the security of 'bulk or link' communications. As the demands on communications media increase, there are greater calls from business, the military and government organisations for the adoption of one single, universal method of communicating data from one centre to another. These data can be a package of voice, text, video, teleprinter or computer data all multiplexed together and transmitted enmasse on one communications link. Combining the various forms of communications becomes ever more attractive as technology, particularly that resulting in increased data rates, makes the techniques financially appealing. Standard data rates of 100 Mbps are common for microwave, leased line and fibre optic links, but as we shall see, not all links can carry high data rates. The military still rely on HF transmission for much of their communications, but inevitably, the data speeds are much lower. Typical bulk transmission equipment offers services from 256 bps to 2 Mbps. As link communication characteristics improve, so the methods of securing them must adapt to meet that challenge. Two environments are discussed in the latter part of the chapter, i.e. a civilian application and a military application, though both have many common features.

8.1 Basic Technology of Link Encryption

The basic configuration of an encrypted link is shown in Figure 8.1. It is made up of a multiplexer as the DTE, a cipher machine and finally a transmitter as the DCE. The multiplexer is responsible for the collection and distribution of a variety of input signals from numerous sources. It converts those signals into digital packages and, in time division multiplexing access (TDMA), arranges those packages into dedicated time slots, before passing the composite data stream to the cipher unit. The time slots can be thought of as logical channels connecting, for example, a local fax machine to a remote machine. The cipher unit carries out stream ciphering of the TDMA signal and forwards the resulting cipher stream to the transmitter, the nature of which will be according to the actual transport medium, e.g. microwave, fibre optic, etc. At the receiver, the cipher stream is deciphered, and the resulting plain TDMA signal is de-multiplexed so that the contents of the dedicated time slots are dispatched to their individual destinations.

8.1.1 Frame Modes

Some consideration must be given to the frame mode of the transmission, and the basic structures for two scenarios are illustrated in Figure 8.2.

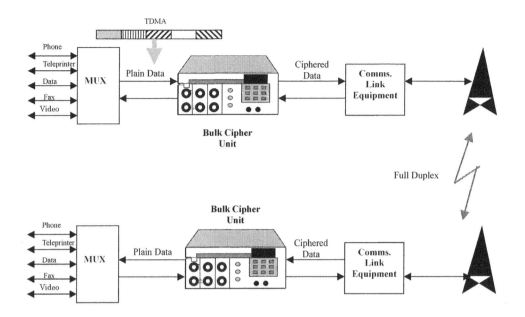

Figure 8.1   The basic structure of a secured communications link

1. A direct link between two stations with no intermediate switching or repeater operation is shown. In this case, all the data transmitted will be ciphered as in the 'unframed/ unframed.' The data are not checked by the data ports.
2. Alternatively, in the case of 'framed/unframed', the frame structure is only checked by the DTE data port. The plain frame includes the frame alignment signal (FAS) and then the whole of the data ciphered for transmission. This mode would be used in networks if no frame structure were used in the secure area of transmission.
3. In the case of 'framed/framed', the frame structure will be checked at both data ports. This mode should be used in switched networks where devices such as routers, multiplexers or repeaters require the frame structure to handle the data. In this case, only the payload will be ciphered. There are variations on this theme where individual time slots in a PCM frame can be treated in isolation for ciphering or plain transmission.

8.2 The Ciphering Process

The encryption process is illustrated in Figure 8.3 and is described as follows.

In transmit mode, plain, digital, multiplexed data, irrespective of its origin, is brought from the DTE into the 'plain' interface where the signal is processed according to the characteristics of the MUX output signal, to match the cipher units internal signal requirements. At this juncture, the preamble part of the synch process precedes the data signal and bypasses the cipher process to be transmitted over the link. The remainder of the synch pattern is used to

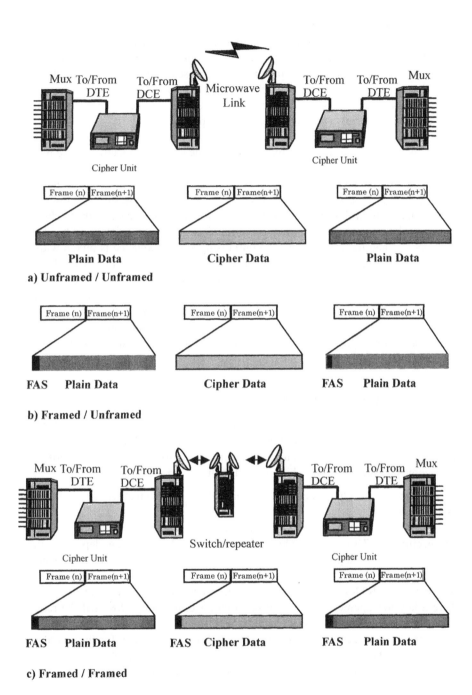

Figure 8.2 Basic frame transmission frame modes

initialise the *TX Key Generator*, and simultaneously, the same pattern follows the preamble but is ciphered before being transmitted. Thereafter follows the actual plain digital TX signal, which is also fed to the cipher unit where *modulo 2 addition* is performed on the data stream by the TX keystream. The cipher stream is then passed to the 'cipher' interface, which modifies the cipher unit's output signal to match the transmitter's input characteristics. The prepared, ciphered TX signal is then transmitted by whichever means is suitable for the link, e.g. radio, microwave, telephone or fibre optic.

In reception, the received signal from the radio transceiver is processed by the DCE and then fed to the 'cipher' interface where, once again, the signal characteristics are matched to the input requirements of the cipher unit. The synch preamble that precedes the data signal is then extracted. As the TX synch pattern was ciphered on the transmit side, it must now be deciphered, by the same key, in the RX synch detector. The resulting de-ciphered synch pattern is then used to initialise the RX key generator. It is clear to see that for synchronisation to be successful, the synch data and the encryption/decryption of the data must be identical in both the receiver and the transmitter. The initialising synch sequence is mixed with the SK value to give matching keystreams at either end of the transmission link. After deciphering, the RX plain signal is once again passed through the 'plain' interface for signal compatibility processing, before being passed to the MUX for demultiplexing. The demultiplexer then distributes the individual packets of signal data to their relevant destinations, whether they be fax, video, phone or data. The system offers full duplex operation and must make provision for a number of a comprehensive array of interfaces to cater for the differing data signal standards.

8.3 Cryptographic Parameters

The cipher system will certainly use a flexible key system such as the three-key system described previously to cope with any timing problems involved in the key change procedures. The ideal machine will have a multi hierarchical key structure for the ciphering process and management administration, which should include:

1. A key assigned to the channel or link, i.e. channel key (CHK) that will be used to generate 'session keys'.
2. A session key, or secret key (SK), which is responsible for the data payload encryption for each specific link. This will include an automatic, timed key change, depending upon the security manager's judgement, but in any case within a defined maximum period of validity.
3. A transport key (KTK) for secure distribution of channel keys and other parameters.
4. Where a key management centre is used, a data storage key (DSK) will be required to secure the database from prying eyes.
5. A key management centre will require default KTKs and DSKs for the initial start-up condition and when a system reset or emergency clear has taken place.
6. If chip cards are to be used for key distribution, the management centre and each cipher unit will need a card write/reader.

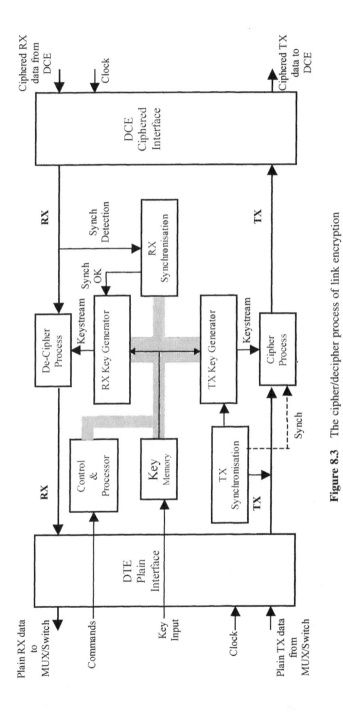

Figure 8.3 The cipher/decipher process of link encryption

8.3.1 Key Agreement

The key agreement between the two stations is carried out to generate a SK for encryption of the data flow within that session. In this manner, a different SK is used every time a resynchronisation is carried out. The SK generation relies upon the two stations sharing a symmetrical channel key identified as the PCHK, i.e. the present channel key, in Figure 8.4. This expression infers that a multikey system is used. The PCHK is mixed with two random numbers that are generated by the transmitting and receiving stations. The random sequences are shared between the two stations and used in conjunction with the symmetric PCHK to provide the key input for the algorithm and keystream generator. The SK output is then used in modulo 2 addition to encrypt the TDMA data in the transmitter and simultaneously decrypt the ciphered data arriving at the receiver.

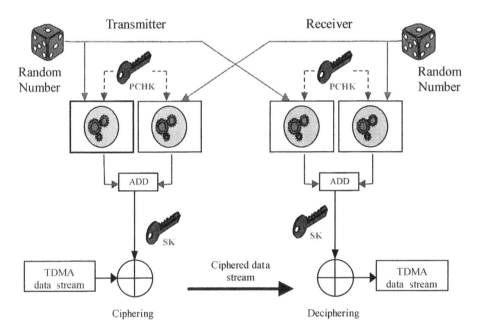

Figure 8.4   Key agreement

8.4 Key and Network Management

8.4.1 Civilian Application

Consider the network 'Eurolink' drawn in Figure 8.5 and its key assignment table, Table 8.1. Each link is encrypted by a different SK that is automatically generated from the PCHK assigned to that link. The SK identity will remain the same for the foreseeable future, but the actual key data will be renewed according to the design of the security manager. The decision on the SK validity period will depend essentially upon the degree of threat and the amount of traffic on that link. At data speeds of up to 100 Mbps, the key stream period is an important issue to consider, and therefore, the SK validity must be limited. To prevent the case whereby the SK is never changed, as is quite often the case in the less well-disciplined climes, a maximum 'timeout' of typically 24 h can be used as a default. However, the security manager should have the freedom to program a SK validity of less than the default period if necessary. It should also be the case that on every occasion that a link fails for some reason, e.g. a lightning strike on a node, an immediate generation of a new SK should be carried out, and a new key agreement is carried out. All key changes should be transparent to the user and seamless as far as the data package transmission is concerned.

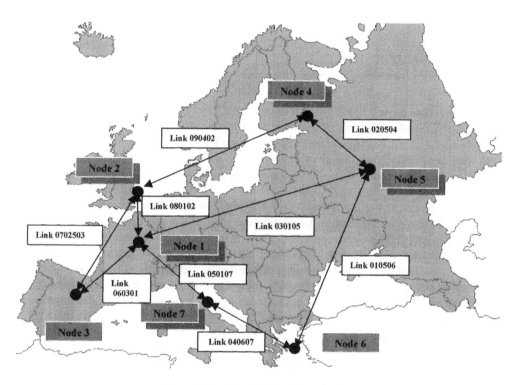

Figure 8.5   'Eurolink' bulk encryption

As the data encryption keys, the SKs, are generated automatically, the network security planning is all about the CHKs. As described previously, the purpose of the CHK is, in conjunction with a random sequence key agreement, to provide for the secure generation of the SKs. Therefore, it is the CHKs that are the entities to be distributed around the network, i.e. they are the fundamental elements of the network security and the main subject of key management, in this instance. Table 8.1 infers that the CHK period might be a weekly event, e.g. week 33 of year 2001 with a spare or next CHK being loaded for the subsequent week, week 34. Again, the use of a multi-key system permits seamless key changes to take place, providing of course that next channel keys (NCHK) are distributed efficiently. Naturally, the distribution of keys over vast geographic distances is not always an easy task and requires a

Table 8.1   Key assignment for 'Eurolink'

Link no.	Node name from	Node name to	SK	Present link key	Future link key
010506	5 Moscow	6 Athens	001	3301	3401
020504	5 Moscow	4 Helsinki	002	3302	3402
030105	1 Paris	5 Moscow	003	3303	3403
040607	6 Athens	7 Rome	004	3304	3404
050107	1 Paris	7 Rome	005	3305	3405
060301	3 Madrid	1 Paris	006	3306	3406
070203	2 London	3 Madrid	007	3307	3407
080102	1 Paris	2 London	008	3308	3408
090402	4 Helsinki	2 London	009	3309	3409

well-planned strategy and its efficient implementation. It might well be the case in 'Eurolink' that the CHK distribution from the network Paris to say Moscow is problematic and that a monthly change of the CHKs is a more realistic target. This is not a major problem and can be dealt with quite easily, even though other links might continue to require a weekly change of CHK. The links are independent of each other, and therefore, the SK key changes are also independent.

In circumstances where a universal key change and distribution of CHKs policy is not possible, the main complication is the distribution of CHKs to individual nodes. Take, for example, Node 5, i.e. Moscow, where perhaps infrastructure problems inhibit a weekly distribution, but where Node 4, Helsinki's link 020402 to London, is considered under threat. The CHK distribution package, probably distributed by 'chip card' or 'fill gun' can be modelled as in Figure 8.6. Moscow needs a monthly card, which must be distributed one week prior to change and, after that change, must be used to load individual keys into each specific link-ciphering machine in preparation for the next change in four weeks' time. The easiest solution for Helsinki is to send a weekly card for the machine ciphering the Link 090402 and every four weeks send a card with the CHKs for both links 090402 and 020504. The links have been listed such that the first two digits represent a unique number in the system, the second two digits represent the originating node ID of the duplex pair, and the third pair of digits represent the destination node ID. How the links are identified, does not really matter, except that they are easily distinguished from one another. In a relatively simple network like 'Eurolink', there are not so many links to deal with, but in practice, they can be many, and unless care is taken to organise identification, chaos may ensue, and troubleshooting problems become a real headache.

As a safety feature, to prevent a link going down in case of a key error, a CHK change over, must only be allowed to take place if, in fact, there is a valid NCHK loaded at each end of the link. In other words, if a key agreement fails, the status quo will be maintained until new keys can be distributed. In the meantime, the link will continue to function with the 'old' PCHK in place. It goes without saying that alarms must be present and give a clear indication to the node operator and the security manager that a 'failed key change' error condition exists.

As shown in Figure 8.7, the standard node will contain one cipher unit per link, and therefore Moscow will have three bulk cipher units, whilst Helsinki has two. It would make life very much easier for the security manager and the node operator, if each cipher unit has a local ID distinguishing it from any other machine in the network. In this case, the unit's ID could be used as an access token in key downloads from a chip card or any other medium. This would ensure that only the security parameters specific to that machine are in fact accepted from the distribution device. It would then be a simpler matter for the node operator to present the courier device at each cipher unit and for that machine to upload only those parameters and keys required. The rest of the card data for other cipher machines would be left undisturbed. This would mean that only one card needs to be sent to each node, and all the local operator needs to do is merely remember their password authorising the card/fill gun action and place the card/gun into each cipher machine's reader slot. The situation described previously required something of a more complex distribution, and it is a rare occasion when one finds a general practice that fits all situations. However, the total independence of each node makes for a flexible, if somewhat cumbersome, solution. Figure 8.6 shows the CHK distribution envisaged for Nodes 4 and 5, Helsinki and Moscow, respectively. It can be seen that the Moscow distribution card carries the NCHKs for all links at that node, and likewise, the Helsinki card carries FLK 3602 for link 020504 (Moscow) and NCHK 3409 for link 090402 (London).

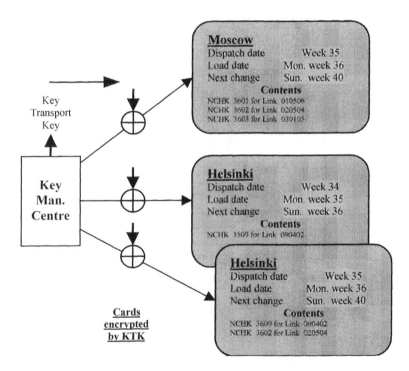

Figure 8.6   Key distribution for Moscow and Helsinki

A second important feature, included in Figures 8.6 and 8.7, is that fact that the distribution cards carrying key data are themselves ciphered by a KTK. The notion of the KTK was introduced in Chapter 2, and here we see a typical application of the function. In this example

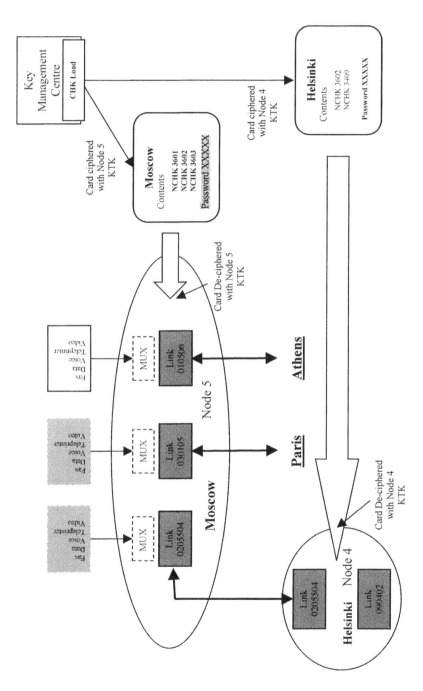

Figure 8.7 CHK distribution by secure chip cards

portrayed, it is applied to chip cards, but it could equally be applied to 'down line loading' or 'over the air' techniques. This procedure opens up a whole new security layer that may not have been obvious before. Apart from the distribution of the CHKs, which are used to generate the SKs, the security manager must take steps to make sure that any key data outside the management or cipher unit environments are not available in plain for anyone to see, hence the use of the KTK. This key is used by the key management centre to cipher the contents of each card so that they may be dispatched with full confidence that their contents will remain confidential. The KTK may use symmetrical encryption, but this is an excellent opportunity to use asymmetrical encryption, which is in fact ideal as a distribution medium as it can provide both confidentiality and authenticity. The latter could also be established by using a MAC and a symmetric key. It must be qualified, that if public key encryption were used, it is likely that only one private/public key KTK pair would be used for ciphering the distributor for all stations. Having a private/public key KTK pair for each link distribution would probably generate excessive overheads, and the situation could be dealt with more easily using individual symmetrical keys for the KTK security. The simplest tactic would be to use the same KTK for all CHK distribution, but then the security of distribution would be degraded.

Figure 8.6 infers that a single KTK is used for encrypting all chip cards used for key distribution, but it would be better security if a different KTK were used for each location. This tactic ensures that there can be no mistake in loading a cipher machine with data from the wrong card as, when uploading a card's contents, the cipher unit must have the same KTK as that which was used to cipher the card at the management centre. More important, as far as security is concerned, is the subject of card loss: if one card is lost or one KTK compromised, the integrity of the rest of the network remains intact. Such a strategy helps to minimise damage to the network in case of some security breach, a very wise policy to employ. The use of KTKs is summarised in Figure 8.8.

Whichever type of encryption is used, the receiving cipher unit will need to have either the same symmetrical KTK as used at the key management centre or the corresponding *public/ private KTK pair*. When considering Figure 8.9, it becomes quite evident that another layer, *the administration layer*, exists behind the message data encryption layer. The new layer is transparent to the fundamental encryption and can be thought of as comprising a series of secure virtual channels. This viewpoint is common in high-security communications systems, as such systems require the virtual channels for security parameter/key distribution. These distribution channels are a basic requirement for the secure distribution of keys, the philosophy being that message encrypting keys should never be distributed over the same channel that they are subsequently intended to protect. For example, if CHKs are to be distributed by 'down-line loading' over say a telephone line that in turn is secured by KTKs, the KTKs should be distributed by a different channel and, preferably, by a different medium. Hence, a mix of telephone and chip card distribution mediums forms an ideal combination. If only chip cards are used for distribution, the KTKs must be carried on a different card to those carrying the CHKs.

The question then arises, 'How can the KTK distribution itself be protected?' The answer is by encrypting the new KTKs by the existing KTK. In the first instance of distribution, or after a machine reset has taken place, a unit's default KTK would be used.

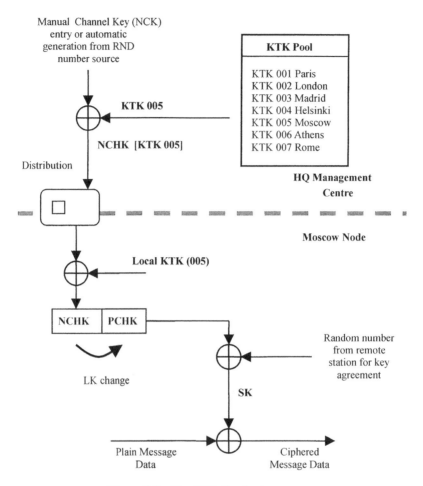

Figure 8.8   The CHK distribution process

8.5 Military Link Security

The demands of the modern battlefield, which now rely heavily upon state-of-the-art weaponry, command and control features, mean that a vast amount of telecommunications support is needed to coordinate fast-moving events. Weapons, logistical data, battlefield images, voice commands and text messages flow continuously between command centres and troop units scattered across the whole scenario. To bring about the successful conclusion of a mission, an increasingly large volume of traffic must be transmitted using broadband technology over radio/microwave relay, fibre optic lines or conventional communications links. Whilst advancing technology strives to satisfy the burden of secure communications, the greater complexity leaves a system more vulnerable to attack, yet it becomes increasingly imperative that a force's communications are protected against the extreme threats that can be expected in times of conflict. This section looks at some of the specific aspects of battlefield link communications security.

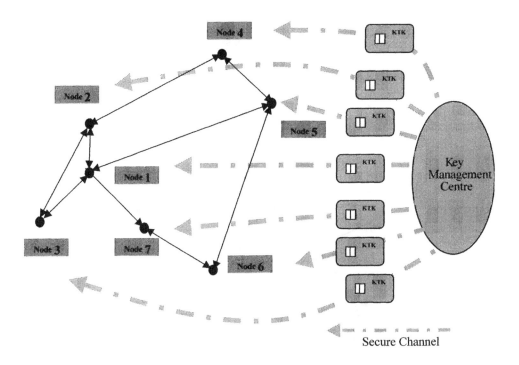

Secure Channel

Figure 8.9   The administration layer of KTK key distribution by secure management channels

8.5.1 Military Topology and Features

Whilst the network topology and security management might not be far removed from that of 'Eurolink', there is no doubt that the hardware components will appear to be radically different. The most obvious difference comes from the requirement that any battlefield equipment has to meet the military standards for mechanical and environmental conditions. In contrast to 'Eurolink', a civil network and therefore of a more semi-permanent nature, military battlefield equipment must be highly mobile and flexible, and the hardware reflects this. It is also very likely that a different transmission medium is used, typically HF radio communication, and that means a difference in data speed. There is also a greater need for the relaying of information between highly mobile forces and the more stationary bases, e.g. HQs, responsible for co-ordinating their operations. In this environment, the 'fixed' structure of 'Eurolink' would be impractical, and 'patch and play' would be very much the order of the day. The introduction of a tactical switch to a link system adds another dimension of flexibility, in that not only can node terminals be fully interconnected, but there will also be the possibility of the switch acting as a hub. This feature will allow access of 10 or more channels for the local insertion of terminal signals, perhaps from a field telephone, into the system backbone. This provides an ideal access point for liaison, or co-ordination officers and is illustrated in Figure 8.10 with the node forming a major role within the model network 'Milink' in Figure 8.11.

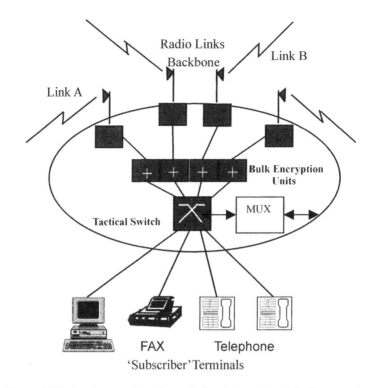

Figure 8.10   Implementation of a tactical switch node with bulk encryption

Figure 8.10 shows a typical military communications node, i.e. the conjunction of four backbone links that may be formed by radio, fibre optic cables, field wire or public networks. Each of the links is encrypted by a dedicated bulk encryption unit, which are connected to the tactical switch. Data arriving from Link A will be deciphered by their encryption unit, and the TDMA channels can be disseminated by the switch and a demultiplexer and then fed to their individual channel destinations. Alternatively, the disseminated channel data from Link A can be combined with the subscriber terminal signals, connected at the node and then encrypted for transmission over the multi-channel link, Link B. Another possibility would be to allow the data from Link A to be directly switched through to Link B, thus merely acting as a relay stage.

The typical, tactical switch will include the following features:

- A switching matrix of 1024 channels
- 10 node subscriber channels each of 256-kbps capacity
- Various interfaces including EUROCOM D/1, for example
- Hardware to meet tactical environment specifications
- Free programming of port/link functions
- Compatible interfacing with encryption devices
- Compatible with other networks, either civilian or military on single-channel or multi-channel connections

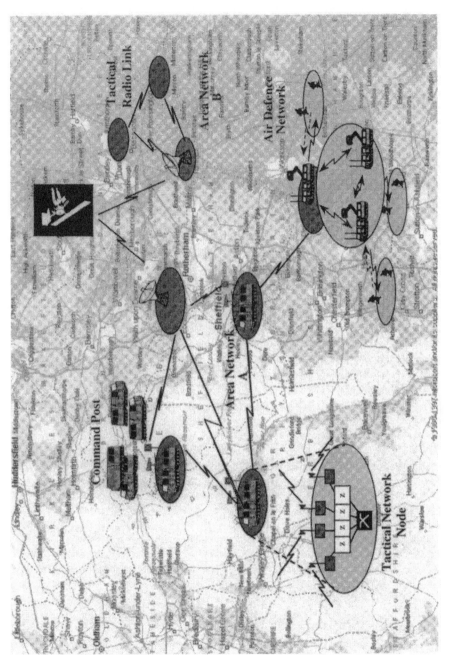

Figure 8.11   The topology of 'Milink'

In a node such as that highlighted above, it makes sense to have central control over individual, modular cipher units for the multi-link encryption. The central control makes for ease of security management, e.g. key loading, change, checking and clearing. A modular approach to the link encryption bestows flexibility for interfacing and trouble shooting and can be combined to give a compact package similar to the arrangement of a vehicle-mounted node, as shown in Figure 8.11.

9

Secure Fax Networks

Despite the revolution in IT communications, especially the massive upsurge in e-mail traffic, and the availability of fax operations on PCs, the facsimile machine is still a popular and convenient method of communicating text and graphic data. Nowhere more so than in military applications, where its simple operation, availability, efficient transmission, reliability and the relatively cheap costs of hardware and transmission are essential. In many parts of the world, the costs of personal computers and the training of staff to use them for communications are prohibitive. Add to that, the necessity of providing IT security as well and the trusty fax machine, although arguably primitive and slow, becomes a favourite alternative for everyday communications.

As far as the purchasing of encrypted faxes is concerned, there are two lines of thought to be considered. They are whether to purchase a fax unit with a fully integrated cipher unit, or to buy fax machines along with 'stand-alone' cipher modules. There are arguments in favour of both options. From a financial point of view, it may be cheaper to buy fully integrated units when initially building a new network, yet to expand or partly replace an existing network, then adding 'stand-alone' encryption devices to existing plain fax machines is the more appealing solution. The latter case allows the network manager the freedom to chose the various facilities offered by different manufacturers from the open market, possibly at a favourable price and similarly to select his preferred encryption from those available. One drawback to the 'stand-alone' approach is the occasional difficulties of compatibility of the encryption unit with the fax machine, especially as time passes and ageing fax machines must be replaced. One could finish up with a whole assortment of devices, all with differing operating procedures, to integrate into the net. A further drawback is that using separate fax and cipher units would require connecting cables between the two and these connections may well introduce a site for TEMPEST attack as they would be carrying plain data. The cables must be well screened to remove this threat or at least some effort should be made to install them in a secure location and as close together as possible to minimise the problem. Similarly, whilst one would expect a secure fax machine to be EMC and TEMPEST proof according to MIL standards, it is quite likely that a standard fax unit bought as an independent machine, will not be built to the same standards, e.g. MIL STD-461/462. Therefore the security of the combined package might be severely compromised by the inclusion of 'radiation leaky, fax machines'.

However, the real headache is that of the network designer in their implementation. Designing the network architecture of an integrated system and controlling its security aspects is one thing, but carrying out the same with a 'stand-alone' system involving various types of fax machine, is a far more complex beast to deal with.

On the other side of the coin, it is a strategic problem for the manufacturer. Should they decide to produce their own integrated unit, bearing in mind the rather limited sales of encrypted fax machines as compared with that of plain, 'standard' machines? Or should they produce a 'stand-alone' cipher unit with all the complexities of compatibility with the whole market range of standard fax machines? Or both?

In the author's experience, budgetary matters usually carry the day and the network manager is left to implement and administer as best they can. This chapter is dedicated to making their task somewhat easier than it might otherwise be.

9.1 Basic Facsimile Technology

The process of converting a page of text or pictures into an electrical signal, transmitting that signal and then reconverting it from the electrical form back into a hardcopy printout at the receiving end, is adequately covered in numerous books on the subject and therefore there is no need to go into great detail here.

It is certainly worth noting though, that the present operating standards as approved by the CCITT are the Group 3 and Group 4 standards described as follows.

Group 3:

- V.27ter/V.29/V.34 up to 21.6 kbps (33.6 kbps, super G3)

    - MMR, MR, MH and JBIG compression of data
    - < 1 min/page (3 s approximately with super G3)

Group 4:

- Digital (ISDN)

    - Compression of data
    - < 5 s/page

In the CCITT Group 3 standard, the digital version of the scanned signal is transmitted at data speeds of 2,400, 4,800, 7,200, 9,600, 14400 bps by the modem. There are new data transmission standards on the market (V.34), which allow higher speeds over the telephone network up to 33,400 bps (super G3) with the speed available for fax transmissions of 21,600 bps. The critical factor deciding the speed of fax transmission in many regions is not the ever improving fax modems, but rather the quality of existing telephone infrastructure. A network designer is far better rewarded by ensuring that their lines of communication are in good order, than in buying superior fax machines which will have to fallback according to the line condition, i.e. typically 9.6 kbps or less.

The main reason for the rapidly improving fax speeds available, is largely the use of compression techniques such as:

- Modified Huffman code (MH)
- Modified read code (MR)
- Modified, modified read code (MMR)
- Joint binary imaging group (JBIG)

9.2 The Basic Operation of an Encrypted Fax Machine

Consider the block diagram in Figure 9.1, which shows the installation of 'stand-alone' cipher units connected to plain fax machines.

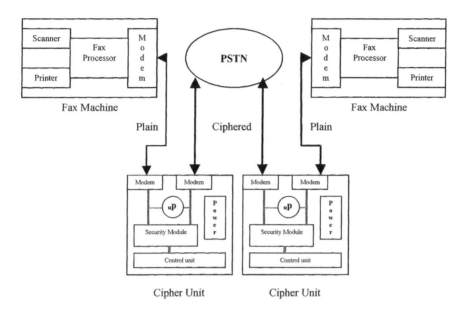

Figure 9.1   Block diagram showing the components of a 'stand-alone' ciphered fax set-up

9.2.1 Fax by Telephone Line

In the transmit mode, once a link has been established by the standard fax protocol described above, the document to be transmitted is reduced to PELS (picture elements) and converted into a stream of binary digits. In order to reduce the transmission time of a document, this binary signal is coded by the MH and MR codes so that in G3 operating mode, a single page document, containing only text, can be transmitted in approximately 45 s. Further compression using MMR code reduces the transmission time to a few seconds, depending upon the complexity and required detail of the document. With separate fax and cipher units, the digitised document must be modulated according to the G3 standard before being passed onto the 'stand-alone' cipher unit. It can be seen in Figure 9.1 that some duplication of processing is apparent in that the cipher unit must now demodulate the fax document signal, back into the digital form, in order that it may be encrypted by modulo addition with cipher unit's keystream, as described previously in Chapter 2. Once encryption has taken place, the cipher stream, resulting from the modulo 2 process is once again modulated and then transmitted along the signal path to the receiver. The duplication of the modulation process might be seen to be inefficient, yet it does confer the advantage that if the cipher unit develops a fault or enters into a blocked mode, it can be bypassed easily by connecting the fax directly to the telephone line. Of course the resulting transmission will then be in plain mode and therefore steps must be taken to ensure that the document is not sensitive in any way. In any case the cipher unit should appear to be a 'transparent' component in both the transmit and receive signal processes, as the characteristics of the ciphered signal are no different to that expected from a plain transmission.

In the receive mode, the signal is demodulated by the cipher unit's modem to extract the cipher stream data before carrying out a 'mirrored' modulo 2 addition with the receiver's key stream to produce the plain data stream. The deciphered data stream is then presented to the 'decompression process' involving the MH and MR processes and then to the printer function of the fax machine which converts the digital signal into PELS and the hardcopy printout is achieved according to the printer type and characteristics.

9.2.2 Fax by Radio

Transmitting faxes by radio follows quite a different process as far as the fax protocol is concerned. First of all a different modem is required, but the main difference is that as it is a simplex operation, no handshake is possible. In this case it is necessary to be able to program the transmitting fax unit in a 'broadcast' mode whereby the fax document will be transmitted whether or not the receiving station is in a condition to receive it. It follows then, that in a broadcast mode, all fax/radio stations tuned to that frequency, would receive the same facsimile providing that each station has the same secret key (SK) selected. Hence the network manager has little control over the traffic on the radio links. Bearing in mind that the exchange of data such as the customer subscriber identity (CSI) is normally part of the handshake, this poses a problem for the key selection mode when an encrypted fax is to be used over radio. Section 9.2.3 describes how this problem may be solved.

9.2.3 The G3 Fax Protocol

Once again, details of the above handshake protocol are adequately described in other available books on the subject, but what is of real interest to the security network manager and hence this chapter is the exchange of identities of the two participating machines. It is this information, which can be used to assign the SKs for that specific fax link, within an automatic fax network. The CSI, which is normally entered manually into the fax machine, rather than the attached cipher unit, can, but need not be, the actual telephone number of the fax station. In fact, as this ID would be displayed on any partner fax in a communication, it is usually preferred that the CSI be a coded or fictitious number so that no unnecessary information is given to the remote station. It is also possible in rare circumstances, such as in a single key network, that this number will be the same in every unit. However, in a normal, multi-key network, using *automatic key selection*, the CSIs must be unique (Figure 9.2).

- CED: called station identification

  - 290 Hz tone

- NSF: non standard facilities

  - Polling pre-set
  - Forwarding pre-set
  - Memory available

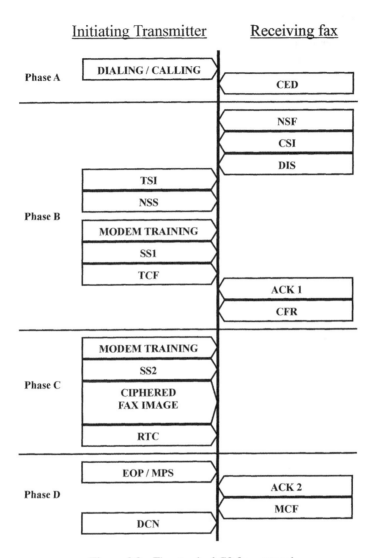

Figure 9.2 The standard G3 fax protocol

- CSI: called subscriber identification

    - Internal telephone number

- DIS: digital identification signal

    - Resolution
    - Paper width
    - Max. paper length
    - Max. speed

- TSI: transmitter subscriber identification

 – Internal telephone number

- Fax image: ciphered picture data
- RTC: return to control
- EOP: end of page
- MPS: multi page signal
- NSS: non-standard set-up

  – Mailbox
  – Multi copy
  – Forwarding
  – Broadcast

- SS1: start sequence 1

  – SK-ID
  – SK-sign

- TCF: training check
- ACK1: acknowledge 1

  – SS1 confirmation

- CFR: confirmation to receive
- SS2: starts sequence 2

  – Commands
  – Data (time)

- ACK2: acknowledge 2

  – SS2 confirmation

- MCF: message confirmation
- DCN: disconnect

9.3 Manual/Automatic Key Selection

With *manual key selection*, the operator must select the SK for each communication link by gaining access to the cipher unit menu and stepping through the pool of SKs to select the correct one. As the name implies, *automatic key selection* is able to automatically select, from the pool of pre-programmed SKs, the key to use for a specific link, by utilising the CSI facility of the fax machine at the remote site. Consider Figure 9.3 and the associated key table in Table 9.1.

The transmitting station receives the CSI from the remote station during the handshake and searches through its own CSI/link/SK table (Table 9.1) to find the correct key. The key signatures of the selected SKs are checked later on in the handshake and these must be the same in each participating unit before cipher fax transmission can take place. If they are not the same, then it will be impossible for the cipher units to synchronise and the fax link will be interrupted.

EASYNET (Multi Key auto net)

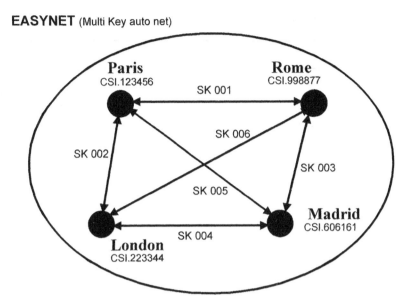

Figure 9.3   A multi key automatic fax network

9.3.1 Multi-key Fax Networks

Hence in 'EASYNET' (Figure 9.3) when Rome wishes to fax London with a ciphered document, Rome calls London and then receives London's CSI 223344. The Rome station searches its link/SK table to find that SK 006 should be used for this transmission and then checks that the signatures are correct in the SS1 part of the handshake (see Figure 9.1). If all criteria are satisfied then the ciphered transmission can go ahead.

This is an example of a *multi-key network*, which uses a different SK for each link. Of course it is possible to use a single key for all the links in the network and this tactic is certainly easier to implement. However, it does present a security problem in that if a single station or link is compromised, then the whole network is compromised. Whereas in a multi-key system such as 'EASYNET,' should the Paris station be lost, SKs 001, 003 and 006 must also be assumed compromised, but the rest of the network can continue operations, secure in the knowledge that Paris never received copies of SKs that it did not use.

Security can be taken a step further by assigning two SKs for each link, i.e. one key for each direction (see Figure 9.4). In this application, an attacker must have both keys for the link in order to gain complete access to the data traffic between the two stations. Naturally there are prices to pay for this extra security and these are that larger networks would require an enormous number of SKs to be generated and distributed and secondly, that the network is

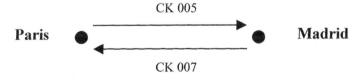

Figure 9.4   Multi-key link with 'unidirectional' SKs

Table 9.1   CSI/link/SK assignments

Paris

Link	Phone	CSI	Secret key
Rome	00396 xxxxx	998877	SK 001
London	0044 yyyyy	223344	SK 002
Madrid	00340 zzzzz	606161	SK 005

London

Link	Phone	CSI	Secret key
Rome	00396 xxxxx	998877	SK 006
Paris	0033 tttttttttt	123456	SK 002
Madrid	00340 zzzzz	606161	SK 004

Rome

Link	Phone	CSI	Secret key
Paris	0033 ttttttttt	123456	SK 001
London	0044 yyyyy	223344	SK 006
Madrid	00340 zzzzz	606161	SK 003

Madrid

Link	Phone	CSI	Secret key
Rome	00396xxxxxx	998877	SK 003
London	0044 yyyyy	223344	SK 004
Paris	0033 ttttttttt	123456	SK 005

much more complicated to manage. Indeed even a slightly larger multi-key network than 'EASYNET,' would require very large numbers of key operations.

For example, the number of SKs required for a 50-station net would be:

$$\frac{50 \times (50 - 1)}{2} = 1225 \text{ SKs}$$

9.3.2 Single Key Fax Networks

An alternative solution might be to give each station one unique key with which to cipher faxes, or to use a random key selection method, where each fax cipher unit carries the whole key range, perhaps 200–500 keys, in its key memory and an SK is selected randomly from this key table for each transmission. The method is even more simplified in that the exchange of CSIs, as in the above model, is not required. What is required though, is that all stations must have exactly the same keys in their memories. *Random key selection* is relatively easy to implement and manage, but has two main drawbacks.

- If one station or link is compromised, then the whole network must be assumed compromised, as each station will contain all of the network keys.
- The network manager has virtually no control over who communicates with whom.

Managers must tailor net security management according to their priorities and in the case of a large network, will have to economise on the use of secret keys. There are several options available to the manager.

9.3.3 Facsimile Transmission Over Radio

As indicated in Section 9.1, transmission of faxes over a radio link has serious repercussions as far as secret key management is concerned. In this case, we cannot rely upon the station IDs to select a suitable key for such a link. Instead, security managers need some way of identifying and programming into the management centre, that this particular link is different to the normal telephone link. The programming must force the transmitting fax station to use a pre-selected SK, which is common to both stations, or indeed, to all participating fax/radio stations and that no further key selection attempts should take place.

9.4 Network Architecture

In practice fax networks are often far larger and much more complex than that in Figure 9.3.

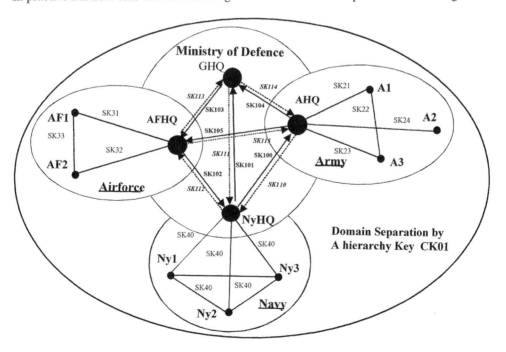

Figure 9.5   DEFNET Topology

The complications are not just a matter of size, i.e. the number of stations and their distribution, but also a matter of security management and in particular the question about access and contact restrictions within the network. Who is allowed to communicate with whom? Furthermore there is the question of nets and subnets to consider.

The model 'DEFNET' in Figure 9.5 raises many of these difficulties and suggests some solutions. Whilst this model is closer to the real situation, there are a multitude of options available to the network manager.

9.4.1 Interesting Features of DEFNET

In this example, the manager has chosen to isolate each force into individual subnets with each of the subnets having an HQ station through which all communications to the ministry and beyond must pass. There is certainly some sense in adopting this feature in that it gives

Table 9.2 Key assignment for DEFNET network

GHQ	CSI	SKID	AHQ	CSI	SKID	NyHQ	CSI	SKID	AFHQ	CSI	SKID
AHQ	12345	114	GHQ	15566	104	AHQ	112345	100	AHQ	12345	115
NyHQ	18994	111	NyHQ	18994	110	GHQ	15566	101	NyHQ	18994	102
AFHQ	12243	103	AFHQ	12243	105	AFHQ	12243	112	GHQ	15566	113

ArmyHQ	CSI	SKID	A1	CSI	SKID	A2	CSI	SKID	A3	CSI	SKID
A1	334455	21	AHQ	12345	21	AHQ	12345	24	AHQ	12345	23
A2	336677	24	A3	338899	22				A1	334455	22
A3	338899	23									

NavyHQ	CSI	SKID	Ny1	CSI	SKID	Ny2	CSI	SKID	Ny3	CSI	SKID
Ny1	26543	40	NyHQ	18994	40	NyHQ	18894	40	NyHQ	18894	40
Ny2	27654	40	Ny2	27654	40	Ny1	26543	40	Ny1	26543	40
Ny3	29876	40	Ny3	29876	40	Ny3	29876	40	Ny2	27654	40

Airforce HQ	CSI	SKID	AF1	CSI	SKID	AF2	CSI	SKID
AF1	44567	31	AFHQ	12243	31	AFHQ	12243	32
AF2	45678	32	AF2	45678	33	AF1	44567	33

each HQ the opportunity to monitor all through traffic, before forwarding it elsewhere. However, the design is rather rigid and readers are invited to expand the network as they wish, perhaps by adding links between say *Ny3* and *A3*, or forming a Tri Force subnet incorporating *AF1, A1* and *Ny1*, for example.

In the army subnet, the manager has chosen to give *A2* only a direct link to the *AHQ*. This infers some special treatment, suggesting perhaps that *A2* might have a degree of priority or alternatively warrant a degree of caution, as it is not allowed to communicate directly with *A1* and *A2* (Table 9.2).

9.4.2 Ministry of Defence Subnet

The Airforce subnet is a normal multi-key subnet with all links permitted, which is also true for the Navy subnet. However, there is a major difference in that the Navy net, perhaps being a large subnet, is only using a single key. As mentioned previously, this idea leaves the subnet less secure and vulnerable to attack, but it does lend itself to economy in SK use. Yet where there are unforeseen difficulties, a single key, so employed, is far more secure than plain communication. It is not recommended to use a single key net, but is included here to show the variety of the alternatives.

The Ministry subnet is a full multi-key net connecting the different HQ stations and therefore it is essential that optimum security be achieved. Bearing this in mind, the manager has decided to implement the option of using 'unidirectional' keys for those links.

DEFNET is by no means a large or complex network, yet when a manger starts to collate all of the network data, to generate and distribute keys to each station, it soon becomes apparent that for anything other than a simple secure fax network, a management centre is an essential tool.

9.5 Key Management and Tools

9.5.1 Key Management of DEFNET

As introduced in Chapter 2, a suitable key management centre must be capable of:

- Accepting, arranging and securely storing data such as station names, telephone numbers, CSIs and equipment information.
- Generating keys and passwords from a random generator source.
- Allowing the setting of operating parameters for each individual station, or groups of stations.
- Providing efficient and secure means of key distribution and other data by means of downline loading, smart cards or by paper hardcopy printouts for manual key loading.

The typical secure *fax management centre* is illustrated in Figure 9.6

9.5.2 Key Generation

The best method in generating and assigning keys for a medium to large networks is to generate several hundred keys at one time, store them in a pool and select individual keys from that pool, to be assigned to each specific link. Thereafter the keys and other data can be downloaded or distributed to the station in question (see Figure 9.7).

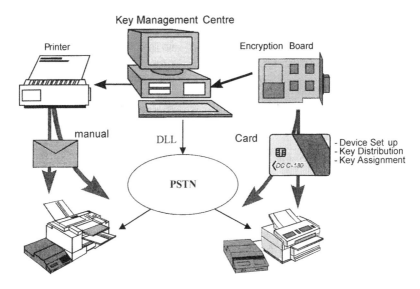

Figure 9.6   Key management centre with distribution tools

Figure 9.7 Illustrates the key/location data package to be downloaded to the AHQ station. For brevity, only the package for the AHQ station is shown but readers are invited to produce their own packages for the remaining stations for DEFNET.

For seamless key changes over the network, it is important to adopt a multi-period key system, especially for a network, such as a fax network, which might well by a world-wide structure. In large networks, particularly those distributed over several time zones, it is unlikely that key changes for the whole network will be carried out flawlessly and therefore some tolerance or flexibility of time must be available to tolerate incomplete key changes. The three key system or period duration keys, using *next*, *present* and *past* key sets, described in

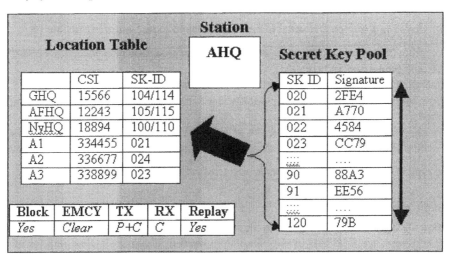

Figure 9.7   Typical key/location/parameter data package for AHQ station

Chapter 2, lends itself to the secure fax network application. Default keys might also be helpful in overcoming key change problems but they too are described in Chapter 2 in some detail.

When using the three key set system, the network manager needs to plan well ahead and the policy of generating a large pool of SKs is most useful when constructing the new key set for the network. The new key set must be generated and distributed to all stations, well before the actual time for the key change command. When this procedure is carried out efficiently, then key change problems will be minimised if not eliminated altogether. The reader's attention is drawn to Chapter 2, for further discussion on key distribution and change operations.

9.5.3 Operating Parameters

The following selection of *operating parameters,* provides a useful management tool for the network manager. The options should include:

- Ciphered transmission only (ideal for a secure, closed network)
- Ciphered and plain transmission (to permit plain communications with non-secure network organisations, e.g. the media or civil bodies)
- Ciphered reception only (to eliminate unwanted fax mail and form a closed secure net)
- Ciphered and plain reception
- Block fax operation, by password (to prevent misuse by unauthorised personnel)
- Emergency clear function (in case of emergency)
- Replay prevention (for authentication purposes)

These too must be packaged for distribution along with security manager and operator passwords that will be used to restrict access to the fax machine to permitted personnel. A two-level password option is very useful in that it allows the informed operator, when necessary, to make less 'cosmetic' changes to machine set-up, without fear of disturbing critical parameters that should only be within the system manager's scope of control. The parameter package maybe different for each station, or common to all.

9.5.4 Key and Parameter Distribution

The key data, locations (CSI) and parameters are ready to be downloaded to the AHQ station, by means of a secured chip card, DLL, and DCD or by paper for manual loading into the cipher unit. In the case of smart card, DCD and DLL operations, it should not be forgotten that data transported in this manner must be protected by the encryption of either medium by a *key transport key* (KTK) so that no plain data is available to anyone who gains illegal possession or access to them. Furthermore, distribution devices such as smart cards should be protected by passwords and if possible, by a *read out limit* and a reference logging of operations, for the managers feedback analysis.

9.6 Fax Over Satellite Links

Until very recently, the transmission of ciphered fax documents over satellite links was almost impossible. However, one or two manufacturers of satellite telephone terminals have recognised the niche requirement and now produce equipment to operate over what is called the global area network (GAN) service M4, with suitable interfaces allowing fax encryption to take place. The terminals are classed as mini-M terminals, designed to operate over Inmarsat M4 with a fax speed of 2.4 kbps.

The original problem was that the usual fax encryption involved the ciphering of the complete document to be transmitted. This included the control/handshake signals such as end of page (EOP) command in Figure 9.2. This is an important signal for the fax transmission and its loss in satellite transmissions, prevented the document from being sent. However, with the new generation of telephone terminals, with their variety of interfaces, the problem has been overcome. With the provision of a 3.1-kHz G3 analogue input interface, the ciphered fax document can be treated as an analogue audio signal just as is the case over the normal analogue telephone line. Then provided that the communication link is of sufficient high quality, as it normally is over satellite channels, then an encrypted fax document can be transmitted both to and from, a fixed ground station to a remote portable unit (see Figure 9.8).

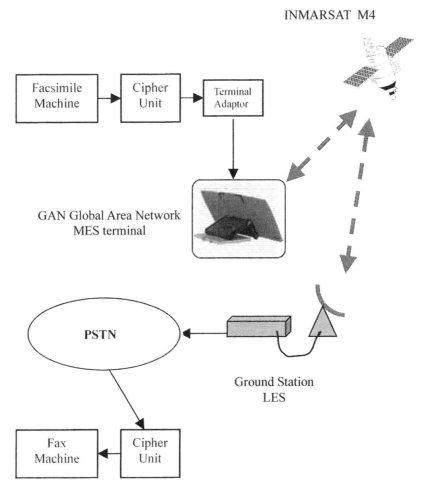

Figure 9.8   Ciphered fax over INMARAST set-up

10

PC Security

We are continually made aware of the security risks to data stored on personal computers and particularly on laptops. Reports are regularly published about laptop computers belonging to security personnel going missing, or successful hacking and virus attacks being made against major companies and government bodies. These serve to illustrate that the risks are at every level, from threats to personal privacy and e-banking as well as those to international politics, commerce and military operations.

The security threats to PC data are many and diverse in nature, as are the solutions available. The computer security market place offers many packages for the protection of PC data, but generally speaking, these measures can be classified as belonging to either software or hardware solutions, or a combination of both. Each has its niche in the market with software solutions such as PGP perhaps, leading a plethora of free/shareware or low-cost solutions for low- to medium-security applications.

Many software products have undoubted qualities, often offering the choice of symmetrical and asymmetrical encryption, 'wipe' programs to render deleted data completely deleted, secure e-mail and access controls. Yet, within the security industry, software encryption solutions are regarded with extreme caution, if not outright suspicion, preferring instead the option of hardware with soft/firmware support. However, the most obvious weaknesses of pure software security solutions, when compared with hardware devices, are that they:

1. are slower in operation
2. can be copied
3. can be manipulated and corrupted
4. have pass-phrases or encryption 'seeds' that are often left unprotected against the enthusiastic investigator, yet they must be remembered by the legitimate user.
5. are more prone to glitches

The serious deficiencies of software security become even more pronounced when one formally considers the threats to PC security as discussed in Chapter 1 and here again listed in Table 10.1, and it can be clearly seen that there is in fact little attraction to software security for the seriously concerned organisation. Therefore, this chapter is dedicated to the subject of hardware protection of PCs.

Hardware encryption is usually available in two forms:

- As 'on-board' devices, installed into an available ISA slot on the computer, supplemented by a smart card or PCMCIA standard module reader for access control. The circuit board

Table 10.1 Security threats to PCs and their solutions

	Threat	Solution
T1	Unauthorised read out of data stored on local hard disk	Stored data on hard disk to be encrypted with strong algorithm. Local storage medium to be removable when not in use by an authentic user
T2	Unauthorised read out of 'deleted data', which often resides in a temporary memory location	Provide a 'wipe' function, which overwrites the residue left after a normal 'delete' operation
T3	Unauthorised read out of data stored on a remote LAN	Provide file encryption by individual file keys. Use a unique session key to protect data sent over a public network. Use removable hard disks
T4	Unauthorised manipulation of data stored on a LAN connected server	Provide an integrity service to protect encrypted files against modification during storage and transmission. Provide an integrity check against the manipulation of the session key transfer protocol
T5	Eavesdropping on the untrusted LAN or public network by wiretapping	Provide confidentiality of files by encryption with individual file keys
T6	Spoofing or masquerading as a genuine member of the trusted network	Provide an integrity check against the manipulation of the session key transfer protocol
T7	Unauthorised manipulation of data during communication over the public network	Provide an integrity service to protect encrypted files against modification during storage and transmission. Provide an integrity check against the manipulation of the session key, transfer protocol
T8	Unauthorised access to the security system	Employ tamper-resistant security modules, which store but never allow unprotected readout of the cryptographic parameters
T9	Unauthorised read out of sensitive cryptographic material such as keys and passwords, in order to gain access to protected information	Ensure that all encryption processes are run within the tamper-resistant module

Table 10.1 (*continued*)

	Threat	Solution
T10	Unauthorised cryptanalysis of the cryptographic parameters and algorithms in order to undermine the security process	Use access control functions such as passwords, that provide a hierarchical access to security functions
T11	Unauthorised manipulation of security equipment in order to gain influence of the algorithm	Emergency clear functions that may also be controlled by time outs
T12	Analysis of stolen security equipment	Factory reset facilities
T13	Unauthorised use of equipment during its transportation to place of use	Refer to Section 10.2.12
T14	A brute-force attack, otherwise known as an exhaustive key search, to establish the contents of key data by trial and error	Provide cryptographic algorithms that use long keys, i.e. > 128 bits, and build a secure security management architecture whereby the key generation, distribution, storage and use and destruction are carried out in a formally protected manner. The provision of custom built algorithm key generation controls are essential
T15	Inefficient security and key management, which would undermine even the most secure hardware and algorithms	Provide comprehensive training, security vetting of management and user personnel. Set up a hierarchical access structure such that users and adminstrators only have access to different security levels on a 'need to know' basis
T16	Analysis of residual plain information	Use comp. emm. equipment to protect against the radiation and conduction of information leaving the system open to EMC and TEMPEST attack
T17	The compromise of information and cryptographic data due to loss or theft of equipment or transfer of security personnel	Provide facilities for a tactical clear, or emergency clearing of keys and a factory reset, which allows for the reset of all security modules and parameters to their delivery states

Table 10.1 (*continued*)

	Threat	Solution
T18	Transmission of message in PLAIN due to loss of keys or incompatibility of keys	Provision of fall-back keys to allow ciphered communication between stations having either lost their programmed keys due to operational errors or having incorrect keys due to poor key management
T19	Access to security equipment during maintenance or repair in a secure or insecure workshop	
T20	Unauthorised intrusion into the PC environment whilst being connected to a public or untrusted network	Make provision for the locking of protected hard-disk partitions, thereby preventing unauthorised access to confidential information

itself would consist of a random-number generator for key generation, a tamperproof key and password memory, a password verification facility and an encryption process for data and keys.

- As adapted modules of the PCMCIA standard, which contain all the encryption, password verification process and access functions as found on the on-board device.

10.1 Security Threats and Risks

The threats to PC security can be classed as falling into six categories, which are further detailed in Table 10.1:

1. Unauthorized access to data
2. Unauthorized manipulation and use of data
3. Loss of data due to deletion or theft
4. Physical damage to the storage medium
5. Physical monitoring (EMC & TEMPEST)
6. Cryptographic attack on security equipment

10.2 Implementation of Solutions

Considering the information in Table 10.1, this section offers practical security mechanisms to minimise, or overcome completely, each of the threats identified.

10.2.1 Unauthorised Read Out of Data Stored on Local Storage Media

This threat has two solutions, encryption of stored data and optionally also removal of the data from the system, which are addressed by the system shown in Figure 10.1. This illustrates how different drives can be configured to store confidential data on secure floppy drives or partitions. The boot partition is not encrypted, to allow easy boot up of the computer. It is possible to modify the boot partition so that the PC will not be able to boot up without access protection to the PC by means of a security module or chip card. This is a complex task and fraught with set-up problems such that it becomes very difficult to implement and use. However, leaving the boot partition unprotected does not compromise the other drives or partitions, which have been designated as secure drives by the system administrator. In fact, the administrator can define and assign all but the boot partition, as they wish. The worked example for the police organisation in Section 10.6 is a good example of such an arrangement.

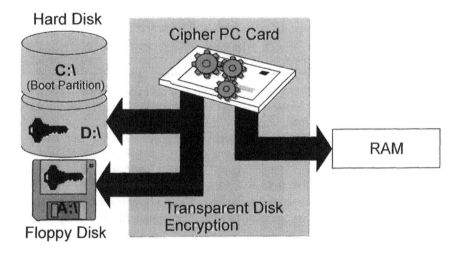

Figure 10.1   Local storage of data on protected partitions and drives.

The data encryption should be carried out using a strong algorithm, but all cryptographic operations should be transparent to the normal user so that file and mail operations are easily carried out and are therefore less prone to operator errors.

The result of this arrangement is that the unauthorised user, without the authentic access rights defined by passwords and personal security modules, will not be able to gain access to, and read files stored on, the secured memory drives and partitions.

10.2.2 Unauthorised Read Out of 'Deleted' Data

Operation of the normal delete function found on PCs does not in fact remove the file data from the storage media whether it is a floppy disk or a hard disk. It merely removes the pointer to the location of that file and so leaves the space occupied by the data within it, open for further storage of other information. For example, if a file is 'deleted' from a floppy disk and then that disk is used to store and transport a new file, the disk will carry not only the new file but also the remnants of the original file, in which case, sensitive information, other than that intended, is available to anyone who has access to the floppy disk. There are many software tools, such as 'undelete' that are available to carry out such retrieval. Of course, if these sensitive data are encrypted, they represent less of a threat but can still add a piece of the jigsaw puzzle for the would-be invader.

The ultimate solution is to use a 'wipe' or 'burn' tool, which writes over the disk space occupied by a 'deleted' file, after which, the original data are completely erased.

10.2.3 Unauthorised Read Out of Data Stored on a Remote LAN

This threat is addressed by the following solutions:

- the provision of file encryption by individual file keys
- the use of a unique *session key* to protect the data transmission over a public network
- the use of removable hard disks

In the case of file encryption, the file to be protected is encrypted by a randomly generated key, i.e. SK, which is good for just that single file (see Figure 10.2). The SK is then itself encrypted by the CMK (customer master key) that is stored on the users card or module by the administrator. The ciphered file is stored wherever desired along with its header, which contains the encrypted SK used for ciphering. Hence, the file and its own SK are stored together, with the security being dependent upon the randomness of the SK and the secrecy of the CMK. It is essential, therefore, that the CMK is stored in an encrypted form on the personal card/module and that it is both transparent and anonymous to the user. Furthermore, the CMK should never be available external to the security module in plain. Group access to files is achieved by the assignment of a common CMK to all members of the group as in the example of the police organisation in Section 10.6.

To decipher the file, all that is required is that the user enters their card or module into the reader provided, selects the file in question and then commands that the deciphering takes place with the plain result to be viewed and stored in a suitable directory, as required. Group

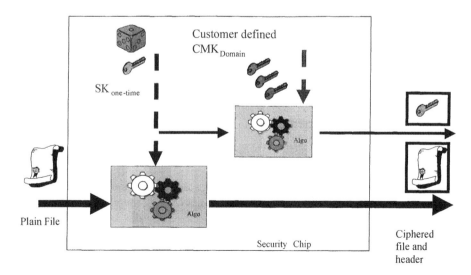

Figure 10.2 File encryption by 'unique key'

access to the selected files is achieved by the administrator loading the group CMK onto each user's card/module. The suitable generation and distribution of a variety of CMKs can be used to construct a hierarchy of file access.

The second solution requires the use of a *session key* that is exclusive to the single transmission of a file. This is achieved in the same way as for file encryption where a unique secret key (SK) is generated from a random number generator and then used to cipher the plain message. In order that the receiver can decipher the message, the SK is itself encrypted by the sender's CMK and attached to the message header, which is transmitted along with the message body. Providing that the receiver has the same CMK, they will be able to carry out a successful deciphering operation.

Therefore, the SK is unique for that session, and it never leaves the security module unprotected.

The use of removable hard disks obviously precludes access to them during periods of inactivity, although it does incur the overhead of a different storage protocol for the hardware.

10.2.4 Unauthorised Manipulation of Data Stored on a LAN

Illegal manipulation of a protected file, when it is to be transmitted for storage, is countered by using a message authentication code (MAC). This procedure generates a cryptographic checksum from the body of the ciphered file and adds it to the body. The MAC is calculated, for example, by using the algorithm in CBC mode such that the checksum is a product of the complete data. This is then encrypted by the message authentication key (MK). The MK is a randomly generated key specific to each single file. It is itself protected by encrypting it with the CMK belonging to that file's creator and attached and transmitted along with the ciphered file for storage within the desired memory location (see Figures 10.3 and 10.4).

At the receiving end, the MK is decrypted by the receiver's CMK, which must be comparable with the sender's CMK and then used to re-compute the checksum of the ciphered file. If the resulting MAC is the same as the original, it is certain that the message or file has not been

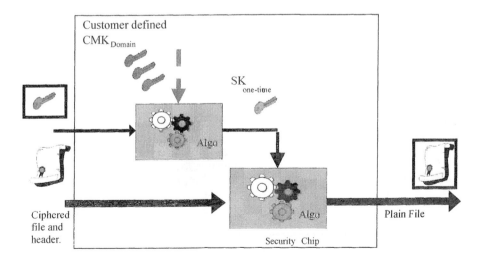

Figure 10.3   Deciphering of a file encrypted by a 'unique key'

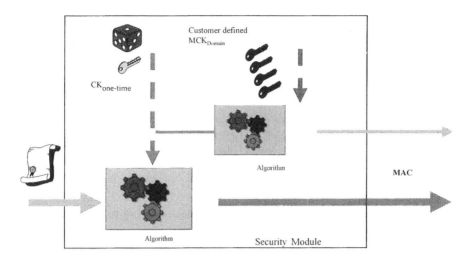

Figure 10.4   Integrity mechanism for a ciphered file

tampered with, i.e. its integrity has been proven. Conversely, any difference in the receiver's MAC and the original MAC will signify that the ciphered file has been manipulated (see Figure 10.5).

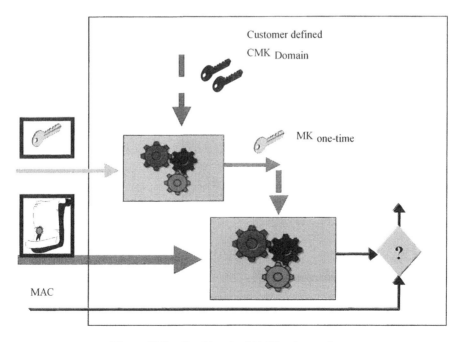

Figure 10.5   Checking the MAC by the receiver

10.2.5 Eavesdropping on an Untrusted LAN or Public Network

The solution to this threat is countered the encryption of files by individual keys, as described in Section 10.2.4.

10.2.6 Spoofing or Masquerading

As discussed previously in Chapter 1, 'spoofing' or masquerading as a member of an organisation brings into question the authenticity of received data. The authenticity of a file transmitted over a LAN or public network is verified by use of a MAC and the key management technique of the security administrator. If a receiver is able to decipher a file with his CMK, then they can be certain that someone having the same key originally ciphered the file. In the case of a group key, the receiver will know that the file came from a member of the same group. Naturally, this is only true if the key management has not been compromised. Measures for protection of key management were discussed in Chapter 2.

10.2.7 Unauthorised Manipulation of Data During Transmission over a Public Network

This is solved by the same techniques as used in protection during transmission on a LAN as described in Section 10.2.4.

10.2.8 Unauthorised Access to/Read Out of/Analysis of/Manipulation of/the Security System

Unauthorised access may be in the form of physical and password/PIN attacks. Physical attack can be addressed by the enclosing of the cipher unit in a 'tamperproof module'. The PC card lends itself to this application, although it is quite easy to implement this in an ISA slot board. As an easily removable component, the PC card, which can contain the complete security package, including access to its functions, can be arranged to run all of the ciphering operations within the module. In this case, no cryptographic data are ever required to leave the protection of the module. The processes are protected against readout, manipulation and addition to the processes by such tools as 'Trojan Horses' by encrypting all of the data in the security module. The heart of the module would be a security chip where the actual ciphering processes take place, controlled by a pre-programmed microprocessor, and once these two building blocks are locked, it is impossible to attack the data stored in them. These components would be supported by a flash-EPROM located external to the security block, and this would contain all the relevant security data suitably protected against unauthorized attacks by cryptographic means. The flash-EPROM would only be written to, and read from, under control of the security measures of the security chip and processor.

Ideally, access to the control mechanisms should be performed in a sealed security chip such as that mentioned above and should typically be based on three role players in the security hierarchy. The passwords of such a hierarchy should be stored and verified within the security chip and hence should be impossible to read out or 'crack'. An error counter with a maximum count of three will prevent an exhaustive guess attack by blocking the password at that level should the readout attempts exceed the maximum. De-blocking of that level of entry should only be available through access at a higher level. Either the 'fatal' or permanent

Strong access architecture in the
security chip makes it impossible for
unauthorized users to use the cipher
keys.

Passwords
prevent misuse

Welded Housing

Logical protection by
encryption with device
individual tamper resistant key

Figure 10.6   Tamperproof PC card

blocking of the module or a very long password combined with a suitable enforced delay between attempts would prevent repeated attempts to access at the highest level.

A typical hierarchy of access might be as follows:

- Security manager: The strategy maker with the ability to unblock all lower access levels, their own access level being protected by a long password, of say 32 hexadecimal characters or a permanent block facility.
- Administrator: Performs the complete management of the cryptographic parameters such as the key management. The administrator's access level should be protected by a six to eight digit password that is monitored by an error counter.
- Operator: Allowed to use the security features to protect the operator's data operations, and their password of six to eight digits should be connected to the error counter.

The control mechanisms should also provide for the inclusion of emergency clearing of equipment and factory resets or blocking functions, to prevent manipulation or analysis when stolen.

10.2.9 'Brute-force' Attack

Otherwise know as an exhaustive key search, the brute-force attack is used to determine the contents of the key data by trial and error. Protection is achieved by providing an algorithm that uses long keys of, say, 128 bits or greater. These keys should be generated either by the security administrator or by the user, as defined under the security policy. In either case it is clear that nobody else should have access or knowledge of the key data. This most certainly includes the manufacturer of the security equipment.

10.2.10 Inefficient Security and Key Management

Ultimately, this is the responsibility of the security administrator, yet it involves every user of the system. Errors and weaknesses of operating practice can be restricted by the appropriate use of access levels, whereby errors, malicious or otherwise, can be limited in their effect on the system. However, the best tactic to limit user errors is to provide comprehensive training on the occasion of the initial implementation of the security policy and thereafter to give periodic top-ups as new innovations or threats emerge. Vetting and testing of operator efficiency will support this idea, as will suitable reprimands applied in the case of blatant security lapses.

10.2.11 Analysis of Residual Plain Information

This is usually present in two forms:

- Electrically radiated information
- Information stored in 'virtual memory'

The first is the result of the radiation properties of the equipment used. PC monitors are notorious for their electrical emanations, and eavesdroppers of the system can easily capture the contents of an unprotected display screen. For the sceptic, one has only to sit in front of a PC monitor for a few hours to suffer the discomfort of sore eyes as a consequence and evidence of the screen's radiations. With quite basic equipment that is available in many a downtown electronic hobby shop, it is possible to 'read' the screen display of a target PC monitor from some distance. The solution is the use of compromising emanation proof and TEMPEST proof hardware. These subjects are discussed in further detail in Chapter 1.

The second source of residual data that is available to the attacker is that of the 'Virtual Memory'. There is a hidden danger to the uninformed computer user, and that is that when a file is being created or edited, temporary copies of those files can be observed to exist for some time afterwards on disk, i.e. a *virtual memory*. This is especially the case on laptop devices where the power supply can be arranged to buffer the computer memory, even during a power-off state.

The problem is overcome by defining a disk partition to be used for 'Virtual Memory Encryption' in the disk encryption settings that should be available to the Administrator. There, it should also be possible to specify the size and location of the paging file. Such settings should override those of the operating system and the TEMP and TMP paths of the user at the time of login. In addition, any application 'default folder' setting for 'temporary files' should be directed to this virtual memory partition.

Due to the operating system's virtual memory architecture, its paging file may still contain sensitive information when shutting down the PC. Temporary files will not be completely erased (wiped), so confidential data may still be physically present on hard-drive partitions after shut down.

10.2.12 The Compromise of Information, Due to Loss or Theft of Equipment or the Transfer of Security Personnel

This threat can be countered by the solution that provides for the reset of cryptographic data back to its factory reset state or delivery state. The reset operation must be carried out before a secured device is to be prepared for shipping or transfer to a different location. On arrival, all the relevant parameters may be re-installed as required.

The loss or transfer of security personnel can be a major threat to a secure system, but this should be borne in mind when establishing the system's security policy and the vetting of personnel. Careful control of access levels will reduce the problem, as will the immediate change of all key data and those passwords, which might be possibly compromised.

10.2.13 The Storage or Transmission of Data in Plain, Due to Loss of Keys or Key Incompatibility

Fallback keys are something of security risk due to the fact that they are most likely to be somewhat common to different users. However, they certainly provide some cover for the occasions when, due to mismanagement at some level, or equipment failure, keys are lost or not distributed, effectively resulting in the plain storage or transmission of data.

The implementation of a fallback key needs to be automatic, hence normally requiring no user action when key availability fails. This action should prevent the storage or transmission of sensitive data in plain.

A note of caution is required here to underline a danger that, in the case of a badly managed system or network, it is possible that the fallback key would be used inadvertently for all transactions due to a key management problem. Naturally, this is a very undesirable state, and steps must be taken to ensure that the user and the system administrator are aware when a fallback key is being used.

10.2.14 Illegal Access to Equipment Under Maintenance or Repair

This threat is countered by the provision of a reset of cryptographic data as described in Section 10.2.12.

10.2.15. Unauthorised Intrusion into the PC Environment Whilst Connected to a Public or Untrusted Network

Without doubt, one of the greatest sources of threat to PC security is from access to an unprotected network. The Internet is the best example of such a case. The golden rule for an organisation wishing to protect its confidential material from Internet based attack is:

- Never access an unprotected network, e.g. the Internet or an untrusted LAN, from a workstation containing sensitive data

The uninitiated or sceptic might like to consider the fact that when a user is connected to the Internet, the remote station is actually downloading data and programs to the target PC. Some of these programs are printing information on the local screen and, in many other cases, exercising controlling functions on the user's computer. It is therefore not difficult to imagine how easy it would be for an attacker to gain direct access to confidential data on the system or the target computer. Even worse is the fact that the attacker can deposit destructive programs, viruses and the like on the unsuspecting PC user, for further attack at a later date. Hence, the PC user, particularly the notebook owner's attention is brought to the rule above. However, the PC and notebook environments are far from ideal, especially from the point of view of user discipline and notebook mobility, and so alternative solutions must be sought. A *disk block* facility is one such solution, the principle of which is shown in Figure 10.7. The blocking is achieved by a control signal to the disk monitor and can be thought of as a low-level mechanical block, preventing access to the disk address bus.

This protective tool enables the user to force designated protected disk partitions to be blocked either when temporarily leaving the user's workstation unattended or when intending to access an unprotected network such as the Internet. For a successful block, the user must close all files and applications that are located on the protected partition and also any disk mapping tools, which might be pointing to these locations. Exercising the disk block tool will then prevent any remote user gaining access to the protected drive partition.

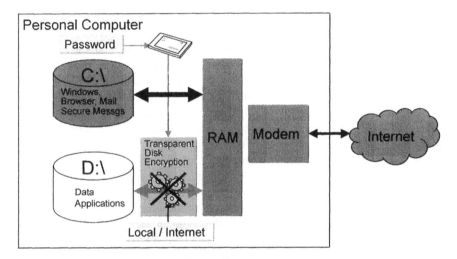

Figure 10.7   An overview of disk block protection

10.3 Access Protection

This strategy guards against unauthorized PC start-up, break functions, unauthorized use, the unintentional or illegal change of data and the deletion of data or programs.

10.3.1 Access Control Systems

These include:

- Use of passwords
- Magnetic stripe cards with a PIN identifier
- Smart cards or chip cards with a PIN identifier
- PCMCIA cards (PC cards) with PIN identifier
- Biometrics using signatures, fingerprints, voice recognition and eye-retina scanning

Passwords are always susceptible to attack, and magnetic stripe information is very limited in capacity and easy to copy. Biometric data are also vulnerable to manipulation, are not always convenient and require extra hardware and therefore increased expense. The fingerprint option of biometric access relies upon the user's finger being scanned and then stored on a chip card to be carried by the intended user. On wishing to gain access to a PC terminal, the user enters their card into a card reader attached to the computer and then places the reference finger on a reader screen. These fingerprint data are then compared with those stored on the chip card, and access to the computer is gained if the two samples match. Should the card be lost or stolen, it is of no use to the invader as the correct finger must be available to be scanned in at the time of presentation of the chip card. Several manufacturers are producing such access devices, e.g. BioSmart ID from 'Idealsoft'.

10.3.2 Access by Chip Card

When wishing to use a protected PC terminal, the user will enter their smart card into a card reader connected to the PC and will be asked to enter their PIN. With a successful PIN operation, a handshake process takes place between the smart card, and the computer's on-board encryption unit automatically authenticates the user. The typical handshake protocol includes not only the PIN but also a check on the PIN error count. Ideally, the card should include an error counter that is incremented every time a wrong PIN is entered. Should this count exceed a predetermined limit, typically three errors, a the card should be blocked until it can be reset by the system administrator. In this manner, brute-force attacks on the PIN can be eliminated. Other factors established in the login handshake are the user's rights to file access. This can be linked with the keys such as the CMK that are actually stored on the card. A further useful innovation, which should be available, is the use of a 'watch list'. The watch list function takes the serial number of the user card and checks this against a black list, which contains the numbers of all rogue cards, as identified by the administrator. On the recognition of a rogue card being entered for authentication, the on-board encryption device will lock the user's card, rendering it useless until a system manager's reset to a default condition can be carried out. After a successful login has been carried out within a certain specified 'timeout' period, various security facilities are then available to the operator according to their profile. This set-up is illustrated in Figure 10.8.

The 'split function' hardware method shown relies upon the security features being split between the user's smart card and the 'partner' hardware installed on the PC including, perhaps, a resident smart card, i.e. the on-board card.

The start-up sequence would be:

1. Company/organisation authentication: do the on-board and user card correspond to each other?
2. Is the user not listed on the watch list?
3. Does the entered PIN correspond with the PIN stored on the user card?
4. The user is authenticated, and keys are freed for use.

10.3.3 Access by PC Cards

The best options are the use of either *chip cards* or *PCMCIA cards*, otherwise known as PC cards, with supporting PINs, or fingerprints. These offer the most cost-efficient methods of

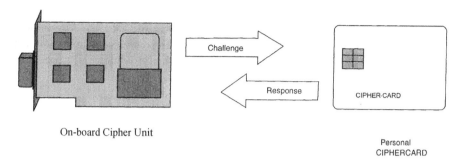

On-board Cipher Unit

Personal
CIPHERCARD

Figure 10.8   Cipher card handshake with on-board device to establish access rights

user identification and authentication. The main advantages of using cards as access controls are:

- With suitable programming, the card can forbid illegal replication and duplication of identity.
- The identification process becomes decentralized and is simply achieved by the possession of the card itself and knowledge of the PIN.
- Depending upon the type of card and its contents, operation of the PC security, whether working in ciphered or plain modes, would be largely transparent to the user.
- Possession of a card control allows flexibility of use in that the user may not be confined to one PC, notepad or workstation. Alternatively, they may indeed be restricted to work on their own station as specified by their security profile.
- Cards can be programmed with a 'timeout', i.e. a defined period of inactivity; thereafter, a re-authentication process is required to be carried out before the user can resume operations.
- All card activities can be logged and analysed at a later date by the network manager.

The features of chip cards and PC cards are discussed in some detail in Chapter 2, but it is important to distinguish here the different characteristics between the two and the way in which they can be used in PC security.

The typical smart cars or chip card has processing facilities and a storage capacity of 16/32 kb with the 64-kb version expected to be released onto the market in the near future. This compares well with the stripe card, which can only carry about 19 bytes. It is this capacity and the limited processing power available that restrict the functionality of this type of card and therefore the role it plays in computer security. Whilst it is able to carry personal IDs, PIN verification procedures and user profiles, the smart card can also be used for key storage and/or distribution. All sensitive data contained on it is encrypted by a transport key KTK. This protection greatly reduces the possibility of readout and copying by any illegal party, as these attacks would only be possible with knowledge of the KTK itself. Therefore, the loss of a card, whilst being a nuisance and a financial loss, would not represent a security catastrophe for the organisation.

The physical structure of the PC card (see Figure 10.6) lends itself to both PC and notepad applications. In these cases, fully integrated PCMCIA carriers are installed, whereas the chip card requires an integral or external card reader, which is certainly not very convenient with notepad use. The capacity and facilities offered by PC cards mean that it can be arranged for the secret keys and other cryptographic data to never leave the card. This gives optimum security for both the access and encryption procedures, which can include a random number generator and strict RED/BLACK separation within a tamperproof encapsulation.

10.3.4 Access by PCMCIA Module

The use of the PCMCIA module as a security module, however, has quite a different application and represents an alternative philosophy of PC security. It can indeed, be used as an access token, but bearing in mind the fact that the PCMCIA module is a highly sophisticated tool with storage and advanced processor facilities, far in excess of those found in the simpler smart card, it is able to play a much greater role.

With this method of hardware encryption, there is no partner 'on-board device', and hence no handshake is carried out. In fact, this module can carry the whole encryption hardware and firmware required to drive it, thereby avoiding the rather complex task of customizing each individual PC station during installation procedures.

As an access token, the user would introduce his module into the station's PCMCIA slot and enter their password accordingly, i.e. as user or system manager, etc. The keys and rights stored on the module will be released for ciphering operations.

10.4 Boot-up Protection by On-board Hardware with Smart Card

It is possible to protect against illegal boot up by encrypting the boot up sector and installing security hardware into a spare ISA slot on a PC and altering the computer's BIOS. Access would be by a smart card with a PIN to combine with the installed hardware. This method gives excellent security, but its implementation tends to be complex, as it requires a major hardware installation along with the supporting bios and memory setting adjustments. These are likely to be different for every computer to be installed, and so the theoretical benefits of this method are seriously undermined, and protecting a large network of computers in this manner can be a tedious, frustrating, expensive and inefficient exercise.

10.5 LAN Security

Most PCs do not operate in isolation but are connected in networks either permanently or temporarily, e.g. LANs, WANs, e-mail systems or the Internet. The more networking possibilities a PC has, the greater the risk of an attack becomes. For each publicly reported instance of an attack, in whatever form, there are substantially many more attacks that go unreported and yet still more that go undetected.

No system can be completely protected from an attack. A dedicated attacker will find a way into virtually any system, with or without inside information. It is the task of PC security to make that process as difficult as possible and to limit the damage that can be done once the system has been penetrated. This involves organisational, physical and logical security measures.

The requirements are different according to the environment in which a PC is operated. The two typical situations are:

- A client/server system with several workstations interconnected by a LAN
- A notebook used during business travel.

10.5.1 LAN Workstation Scenario

Figure 10.9 shows a system configuration with basic security protection.

The principal security element is a firewall between the main LAN and the homepage server. The firewall is configured by a number of rules to prevent direct access to the LAN from an external source but would provide loopholes that could be exploited when workstations on the main LAN access the Internet. In addition, once the firewall is breached, an attacker would have access to all the data available on the main LAN.

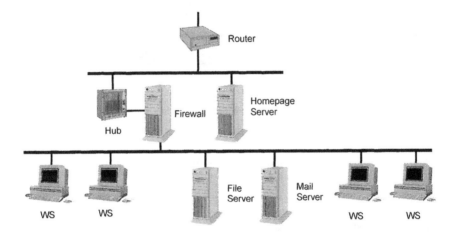

Figure 10.9 Basic protection in a client/server system, scenario

This configuration can be transformed into a much more secure environment by some organisational restructuring and a few additional components.

In the configuration shown in Figure 10.10, separate workstations are provided for Internet access on the same, unprotected LAN as the homepage server. This ensures that no access to the main LAN can be gained while browsing the Internet.

Secondly, the main LAN is separated into two parts: one for normal users and one for users handling confidential data. The two LANs are separated by a firewall to prevent access to the confidential part by normal users. The confidential part is also provided with its own file and mail servers where additional security measures, including encryption, can be implemented that are not required by normal users.

Such a configuration would also be supported by general security measures, such as:

- Restricted access to the principal components of the LAN infrastructure (servers, firewalls, hubs, etc.)
- Allocation of all users to defined user groups with access restricted to only those programs and data that they essentially need

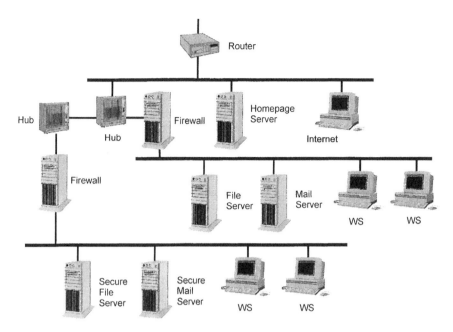

Figure 10.10 Enhanced LAN protection

- A policy requiring users to change passwords regularly and to use password-protected screensavers and/or workstation locking functions
- Encouragement of users to store data on the LAN rather than on the local hard disk
- Virus software to scan data and mail
- A data backup system covering daily backups of the past 30 days and monthly/yearly archives
- Storage of highly confidential data on removable hard disks that are used only on standalone workstations (i.e. without LAN access)

10.5.2 Business Trip Notebook Scenario

In this scenario, the main threat is not from a remote attack, although this is still a major consideration, but from unauthorised local access or theft and from monitoring.

There will inevitably be times when a notebook is left unattended during a business trip, and without access and data protection, the complete contents of the hard disk could be copied without the owner even being aware it has occurred. Also, when working in a foreign environment, the user must rely on local facilities for power, communication and sundry office services, which may be subject to monitoring.

The best way to counter the threat of unauthorised access is to have some form of small token, like a smart card or PCMCIA card, that is required to be present and will request a PIN or password before granting access to the operating system. The token can be removed and kept in a pocket when the notebook is not in use, thereby rendering the notebook inoperable. As a backup to this measure, in case the token is either obtained or duplicated or the token mechanism is bypassed, the confidential data on the hard disk should be encrypted and the encryption mechanism protected as previously discussed.

If both the above measures are implemented, in the event that the notebook is stolen, you can at least be reassured that the confidential data will remain protected and, assuming that you have a backup of these important files, the loss can be fairly quickly recovered.

As far as monitoring is concerned, it is best always to be suspicious and use a policy to safeguard against such attacks. The policy should include:

- Never let anyone else use your notebook; even if you feel you can trust them and think watching them while they use it is a sufficient safeguard
- Never install programs or files during the trip unless you are absolutely certain of their authenticity
- Always use battery power for the notebook; take spares and only recharge where you are confident that monitoring is unlikely to occur
- Always encrypt confidential data when communicating, whether the data forms part of the communication or just resides on the notebook
- Avoid printing and use of copier machines whenever possible; either bring your own or use them only in trusted places.
- Always log out, lock or invoke a password-protected screensaver when called away while operating, even if you are called away for only a few moments.

10.6 Model Application of PC Security

Consider the structure of the police organisation shown in Figure 10.11

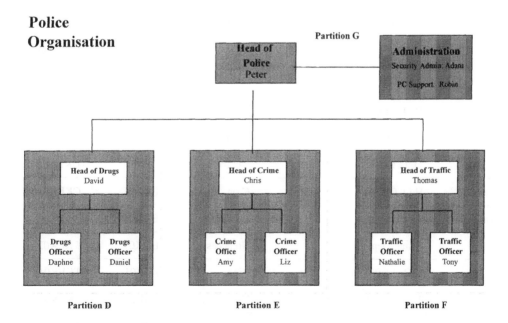

Figure 10.11   Model of PC security organisation–police

The administrator, in collaboration with Peter must:

- Formulate a general security policy for the organisation
- Define all users and departments, taking into consideration the user groups
- Define the security profiles for each member or group
- Define the access rights for each user or group
- Define the file and disk structure for each group
- Define all machines used in the organisation
- Define keys, key groups and key data
- Provide a means and protocol for backup of data
- Provide for key distribution and changes

The suggested profiles are described in Table 10.2.
We can see from the information tabulated in Table 10.2 that Peter, being the 'Head Of

Table 10.2 Security profiles for the police

User	User group	Disk part	Floppy disk	Disk key	KEKs
Adam	Global	All + G	Pl. & Ciph.	All	AKEK01
					AKEK01/02
					DKEK01/02
					CKEK01/02
					TKEK01/02
Robin	Global	All + G	Pl. & Ciph.	All	AKEK02
Peter	Global	All + G	Pl. & Ciph.	All	PKEK01
					AKEK01/02
					DKEK01/02
					CKEK01/02
					TKEK01/02
David	H.O. Drugs	D	Ciph.	DDK01	DKEK01/02
Daphne	Drug Dept	D	No access	DDK01	DKEK02
Daniel	Drug Dept	D	No access	DDK01	DKEK02
Chris	H.O. Crime	E	Ciph.	EDK01	CKEK01/02/
Amy	Crime Dept	E	No access	EDK01	CKEK02
Liz	Crime Dept	E	No access	EDK01	CKEK02
Thomas	H.O. Traffic	F	Ciph.	FDK01	TKEK01/02
Nathalie	Traffic Dept	F	No access	FDK01	TKEK02
Tony	Traffic Dept	F	No access	FDK01	TKEK02

Table 10.3   Key data for police

Key group	Key name	Key data	Sign.	Valid
Global	AKEK01	A489 263E 09B1 668C 2750 BB8F 2291 EC77	A224	01/2002
	02	D558 8924 ED90 BB74 22AC EF49 7721 6544	76FE	01/2002
	PKEK01	B445 CA87 B990 066D ECC2 1687 9883 ABD5	41AA	06/2002
	GDK01	2235 0369 AC45 6709 C421 D567 3301 BE46	33EC	01/2003
Drugs	DKEK01	7459 2235 5961 6622 2546 AB84 129A CC78	009A	09/2002
	02	EED8 0824 E340 CC74 272C 1239 EE21 6789	880B	09/2002
	DDK01	1050 EA46 6900 4582 4792 4688 ACD4 BBCA	EE35	01/2003
Crime	CKEK01	CD88 789E 09C3 748C 4550 EE1F 7091 6412	E557	05/2002
	02	A557 8934 ECC0 6B79 890C AC49 7891 6533	7705	05/2002
	EDK01	1044 BA88 9043 12242 CDE2 FF87 BCA2 1001	203A	01/2003
Traffic	TKEK01	ACCD 263E 8900 9B74 2166 85BB 8F1E 4657	47AD	04/2002
	02	6700 8924 D90B 9B72 29AC EF49 7972 782A	AF54	04/2002
	FDK01	3900 5675 ADF2 10BC 3367 8932 A580 CF21	098C	01/2003

Police', has access to all files on all network drives and has free access to floppy disks. Adam and Robin have their own admin keys and free access to the floppy drives so that they are free to update software and perform troubleshooting functions. Being the security administrator and therefore responsible for the key distribution, Adam has all the keys. He will have the task of defining and distributing the relevant DKs and customer master keys (CMKs) for each individual user.

Each head of department has only ciphered access to floppy disk, his own CMK and those of his charges, e.g. Daphne and Daniel, so that David has access to all the files in his department, which are stored on the Dugs network drive. Daphne and David have pooled files under their key DCMK02. The disk keys, e.g. DDK01, protect each user's local partitions. All members of a department have a common key to share information, e.g. DCMK02, but there is no 'cross-departmental access' to other network drives, except for the global (administration) group, and even here, Robin cannot decipher files.

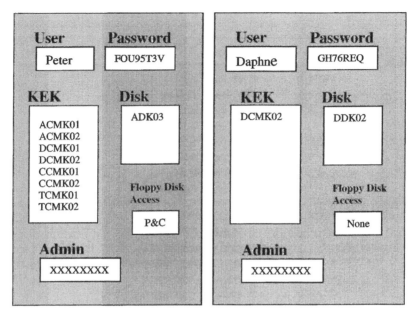

Figure 10.12   Key distribution data for Peter and Daphne

Table 10.3 shows the key data for the whole organisation, though in reality, we would certainly require that the actual key data not be visible to anybody other than possibly Adam, who would be responsible for their generation and distribution. There is a strong argument and a good case for the keys never being displayed at all. A good key management tool would allow the random generation of keys but no means to display anything other than their signatures to facilitate checking out key distribution and change problems.

Key signatures were discussed in Chapter 2. An important parameter in key operations is the validity of keys. It is the admin manager's task to plan ahead for key changes and to carry them out seamlessly and transparently as far as the normal users are concerned. One of the admin manager's problems is how to cope with users' files or floppy disks,

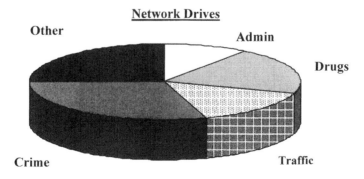

Figure 10.13   Partitions for the police PC security

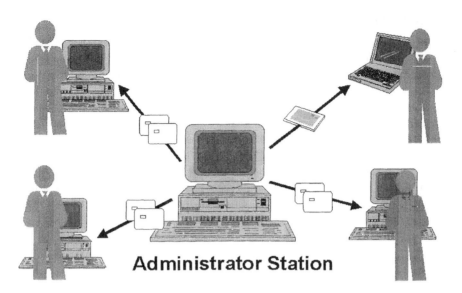

Figure 10.14 Key distribution from a management centre

which are ciphered with 'out of date' keys. A straightforward key change for each department would render its existing files and partitions inaccessible from that point on. One solution would be to give a 'long-term key' to each head of department so that they would be able to decipher these files and then re-cipher them with an updated key. An alternative solution would be to retain old keys just for deciphering old files and then use new keys to re-cipher them. When all old files and disks have been reciphered, the old keys can be deleted.

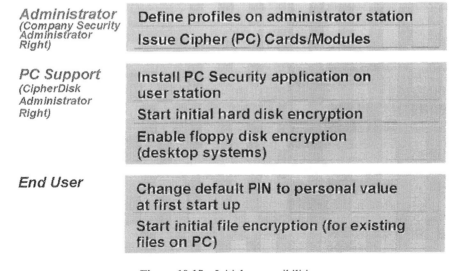

Figure 10.15 Initial responsibilities

10.7 System Administration

Figure 10.12 shows the typical security distribution 'packets' for users. For brevity and contrast, only the data for Peter and Daphne are shown. Once again, a good key management centre would allow the manager to assemble such packets for their users and then load them into smart cards or PCMCIA modules for distribution. Each card/module should be protected by the appropriate user password and password error counter and should also contain one level of admin password, to enable the security manager to unblock the inevitable, but hopefully only occasional, blocked cards, due to password abuse. Such a distribution is suggested by Figures 10.13 and 14.

Typically, a security management team would consist of a security administrator and a network manager. The latter would be more concerned with the installation and hardware tasks, whereas the security administrator would be focussed on the key management and access rights of the system. These responsibilities are illustrated in Figure 10.15

11

Secure E-mail

The United States and its allies are believed to routinely monitor e-mail messages as part of its *Echelon* project, searching for specific words or phrases. This is not just for counter-terrorism reasons but also to facilitate combat against industrial espionage and to carry out political eavesdropping. It is not just national agencies that carry out extensive mail monitoring. There is, in fact a thriving business in providing commercial and criminal elements with the information that they seek on organisations and individuals. The loss of reputation and money by such revelations can be huge if, for example, data containing contracts information or negotiations, financial information or results, personal files and political messages are revealed.

The use of e-mail is now so commonplace that many people, especially the younger generation, have forgotten, or have never learned, how to write letters, formal or otherwise, using pen and paper.

Its popularity is understandable when you consider the advantages that it presents over ordinary postal services:

- Immediacy: messages can be received seconds after they are sent
- Availability: messages can be sent or received from anywhere in the world, even when travelling
- Flexibility: messages can contain attachments of programs and data files
- Adaptability: although the Internet is the most popular means of communication, other forms such as LAN/WAN, radio, or satellite communications are also widely used.

11.1 The E-mail Scenario

Essentially, all that is needed to start sending e-mail is an address and a means of communication. Numerous applications are available for processing the messages, such as Microsoft Exchange, Microsoft Outlook, Lotus Notes, etc.

The address usually specifies the user account at a 'post office', which is where messages are deposited for collection. The post office (Figures 11.1 and 11.2) may be located on an individual PC, an organisation's mail server, or an Internet provider's mail server.

The post office forms the heart of the mailing system as it controls the distribution of mail between its users. User names, passwords, mailboxes and access rights are defined at the administrator level. This enables access to the system to be restricted to defined users only. The administrator can also determine with whom each user can exchange mail. And yet, because of its familiarity and ease of use, what is very often overlooked, especially by the

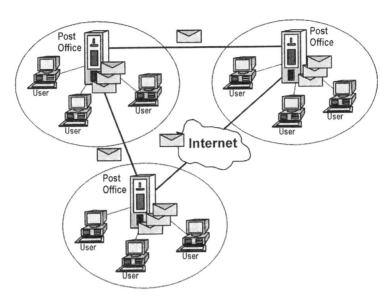

Figure 11.1   The general e-mail network scenario

user, is the security aspect. Confidential information that a user would take great care over when sending by normal postal services is frequently sent by e-mail without a second thought.

For example:

- Sales figures
- Contract offers
- Personal and financial details about oneself or others
- Proprietary technical details or software
- Travel or shipping arrangements

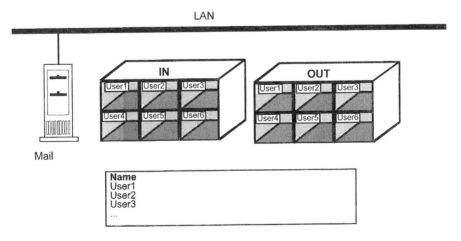

Figure 11.2   Representation of a 'post office'

and many other potentially damaging sources of information are often sent unsecured, even when a secure e-mail system is available.

11.2 Threats

The main threats to e-mail systems (see Figure 11.3) are:

- Information disclosure during transport
- Modification of messages during transport
- Replay of recorded messages
- Masquerade–the sending of messages pertaining to be from someone else
- Spoofing–insertion of false messages in the system

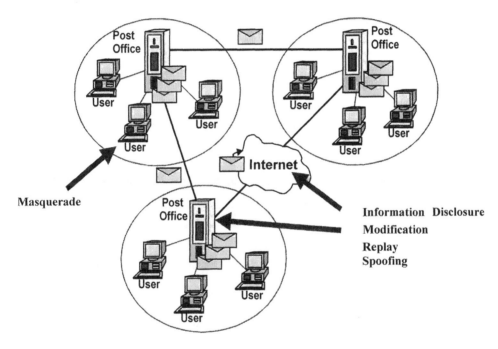

Figure 11.3   Threats to e-mail security

11.2.1 Information Disclosure

Information transmitted in an e-mail may be read by persons other than the selected recipient. There are a number of tools available that can be used to gain access to, capture and display the message data carried by an e-mail. Figure 11.4 shows the mode of connection to capture information from a genuine LAN message. A screen shot of such a capture is shown in Figure 11.5.

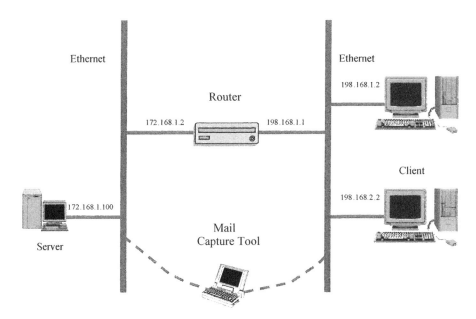

Figure 11.4   Example of a connection to capture mail

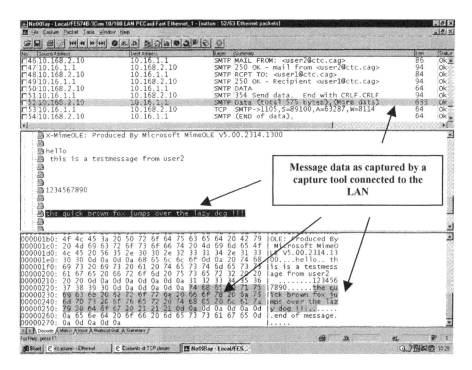

Figure 11.5   Information disclosure by mail capture tools

11.2.2 Modification of Messages

The contents of an e-mail may be modified during transport or storage. The whole message, the attachment or just a few characters of a message may be replaced. This can lead to ruthless fraudulent attack and, in the least, can be the source of misunderstandings, loss of reputation or simple generation of confusion. To overcome this problem, information must be protected in a way that the integrity is guaranteed, and any modification of the message must be disclosed to its genuine owners.

11.2.3 Replay Attack

A message may be recorded on the network and re-sent to other recipients. This may also lead to loss, confusion or damage to an individual's or organisation's reputation.

It is very important that any messages sent without protection do not contain any sensitive information, which can later be used against the user or to damage his/her reputation. All other information should be protected in such a way that any wrong recipients cannot read the mail.

11.2.4 Masquerading

It is possible that a message could be sent by one person using another's name. It is not a difficult task to create a mail and change the address of the sender, in order to hide the real source of the message. This can be done by an attacker who wishes to hide their criminal intent, or to inflict damage upon the purported sender's reputation. For example, by the sending of hate mail or mail with racist or sexual content, under another's name, the attacker can easily bring their target of the attack into disrepute and even instigate legal proceedings being taken against them.

11.2.5 Spoofing

False messages may be inserted into another user's e-mail system. The purpose may be a fraudulent attack, or to create misunderstandings or confusion, or to inflict damage upon a user's reputation. Such messages can be inserted from within a local LAN, or from the external environment, using 'Trojan Horses'.

11.2.6 Denial of Service Attack

This is one of the most common attacks. The idea behind a so-called denial of service (DoS attack) is to put a mail system out of order by overloading it with mail shots. This method of attack is normally carried out by using Trojan horses or viruses, sent to the user in the contents of an e-mail. The e-mail entity will automatically forward itself to all the destinations that are stored in the victim's address book. This attack will continue by the virus propagating itself throughout a system and subsequently will infect one organisation after another. Some very well-known examples of this kind are the 'I Love You' and 'Internet Worm' viruses.

Another way of disturbing a target system is by blocking user accounts. In certain systems it can easily be achieved by repeatedly entering a wrong password in a challenge/response log-in.

11.3 Type and Motivation of Attackers

There are countless people and numerous organisations that are motivated to attack e-mail systems and messages:

- Competitors have financial advantages when gaining knowledge about contracts, researches and product development. They may also engage actively in disturbing the target system by using spoofing, masquerading or modification of messages.
- Information theft will provide a financial advantage to an agent dealing with stolen information.
- Spies are interested in any information that can give his/her employer (country or company), a political, strategic or economical advantage. They may also destroy the reputation of a competing organisation, company or person, using the information captured, or by spoofing, masquerading or modification of messages.
- Disenchanted employees, unsatisfied or curious users gain satisfaction from the possession of sensitive information and may wish to use it to destroy the reputation of their own organisation or certain employees.
- Joy riders are people, who gain satisfaction from the success in getting information by illicit means. They also experience a sensation of power by being able to control systems or put them out of order by using DoS attacks or to destroy the information by using viruses.

11.4 Methods of Attack

There are many different ways for an attacker to obtain or modify the content of an e-mail or to spoof and masquerade them. Here are just a few examples:

- IT personnel: These have access to the servers, where the e-mails are stored. From here, they can access the data so that they may be analysed, copied, printed or modified. It is also easy for the IT personnel to gain access to the e-mail account and send mails in a user's name.
- Server/PC maintenance: Systems sent for repair may still contain sensitive information, which can be analysed by these maintenance teams.
- Server backups: If someone has access to the backups from a server, it is very easy to copy and analyse them. Additionally, a backup normally contains all the messages that are stored on that server.
- Unauthorised user: Unauthorised users may try to gain access, spoof or masquerade the e-mailers logging into the server by using the account and the password of another user.
- Network capturing: All information sent between the clients and the servers over the LAN or WAN may be captured on the network. This may be either the message itself or username and the password to access the e-mail server. It is also possible to spoof or masquerade e-mails using this technique.

- Viruses and Trojan horses: e-mails may be forwarded automatically to a certain recipient or to all recipients stored in the user's address book. This may disclose or destroy information, or put a system out of order by overloading it. The captured addresses may be used as a source for future direct attacks or to determine the members making up a particular network. Search machines may also be used to identify and compile lists of addresses.

11.5 Countermeasures

Countering the threats detailed above requires both logical and organisational means. The logical elements are:

- Cryptography: to encrypt the information so that illegal observers are unable to view it.
- Authentication: to prevent tampering.
- Digital signatures: to verify the message source.
- Virus scanning by either the client or e-mail servers.
- DoS attack can be countered by designing the network so that if a service becomes flooded by a DoS attack, the remaining services remain available. It is very difficult to protect against a DoS attack in an open system, but it is quite easy to trace the source machine of the attack, though this might still not reveal the perpetrator.

These elements will serve to protect the messages, but they also require organisational means to ensure they are available to the people who need them and are used when they are required (see Figure 11.6). In this respect, the following must be considered:

- Policy: what information needs to be protected
- Structure: how the defined policy is to be implemented
- Personnel: who is to have access to the defined structure
- Training: to ensure that the personnel know how to use the defined structure properly and to its best advantage, and are fully conversant with the defined policy.

The policy to be implemented could range from all external messages having to be encrypted, to only certain specified message content to particular locations needing protec-

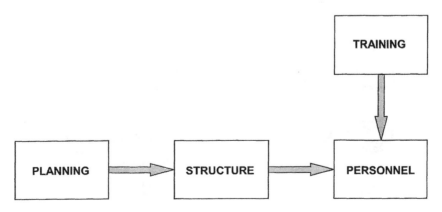

Figure 11.6   The relationship of organisation elements

tion. The actual requirements will vary from organisation to organisation according to the type of business being transacted.

Having determined the policy, the next step is to determine the means by which the policy will be implemented. Here, the structure of the system hardware and software that is required to carry out the policy must be determined and evaluated to ensure that it provides the necessary security. This may involve setting up a separate secure e-mail system rather than incorporating it into an existing system.

Finally, the personnel who will administer and operate the system need to be appointed and fully trained in its use, for there is no point in having a system if the operators cannot or are reluctant to use it, or the administrators fail to make the system workable for the operators.

Returning to the logical means, it has to be decided which form of encryption, symmetrical ciphering or asymmetrical ciphering, is best suited to the network in which it is to be used.

Asymmetric ciphering (see Figure 11.7) operates using a pair of related keys, termed public key and private key, each having a different value. The public key is made freely available to anyone who needs it, while the private key is kept secret by the user. When User A sends a message, they use the public key of the recipient (User B) to cipher it. The recipient of the message (User B) then uses their private key to decipher the message. When replying to the message, User B uses the public key of User A to cipher the message, and User A will use their private key to decipher it.

As this type of ciphering cannot on its own determine the source of a message, it is usually combined with a digital signature to identify the sender. The digital signature is formed by performing a hash function on the plain message and ciphering the result with the user's private key (see Chapter 2). The digital signature is then attached to the message, usually with the user's public key to enable a reply. The recipient can then use the digital signature and associated public key to verify their validity with the independent certification authority, which issued these data.

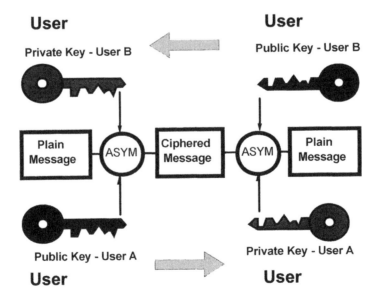

Figure 11.7   The asymmetric ciphering process

This type of ciphering is best suited to 'open' networks where participants may join and leave at will, and key management is to be entrusted to an independent certification authority.

Symmetric ciphering, as described in Chapter 2, operates by both sender and recipient being in possession of the same key. This type of ciphering is best suited to 'closed' networks comprising various user groups (see Figure 11.8), particularly where the organisation is structured hierarchically.

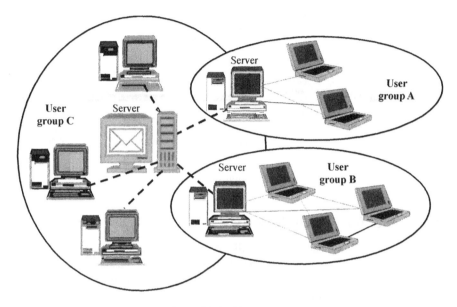

Figure 11.8   User groups

Each user group will have their own common keys so that members of user group A, for example, can communicate in isolation from the other groups and vice versa. With the usual abundance of key availability, individuals within a group will be able to communicate in cipher mode with individuals from another group by both parties sharing a common secret inter-group key. Conversely, individuals without such a key will be able to communicate in cipher only within their user group. In this manner, a whole hierarchy of access and links can be constructed by the use of a well-defined key structure. Chapter 9 describes some typical network structures that are common with this chapter.

Where messages may be addressed to multiple users, it is typical to use a hybrid system of encryption that combines the benefits of both symmetric and asymmetric ciphering (see Figure 11.9). First, the message is symmetrically enciphered using a random session key (as described for file encryption in Chapter 10). The session key is then encrypted using the asymmetric scheme and attached to the message. In this way, the message is only ciphered once, but the session key is ciphered with the appropriate public key of each of the recipients.

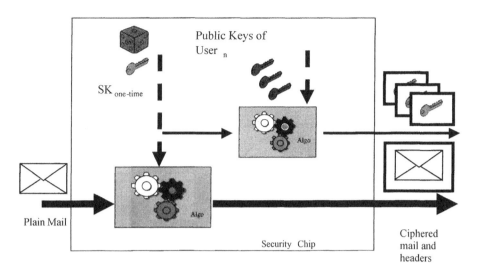

Figure 11.9   Hybrid encryption scheme for multiple recipients

11.6 Guidelines for E-mail Security

The advantages of communications by e-mail are obvious. Its simplicity of operation, immediacy and fast dissemination have tremendous implications for business, politics, defence,
security services and, in fact, every walk of life. The benefits of e-mail are plain for all to see,
but as we have seen in other chapters of this book, in the newspapers, news bulletins and
television, as a communications medium, it is fallible in many ways. We, at this moment, are
concerned about security, and as such, much consideration must be given to the establishing
of controls in order to reduce associated risks to the business and security worlds as they
exploit e-mail. An organisation;s policy, as far as e-mail communications are concerned,
should include the following directives:

- All messages and attachments should be ciphered for transmission by e-mail and the
  storage on mail servers
- E-mail administrators should not be able to view the plain text of their user's messages
- Users should be encouraged to protect their e-mail domain by passwords
- The e-mail addresses of large groups of users should be protected and should not be
  generally available by default to all users in the organisation
- The management team must be prepared to provide alternative services and react promptly
  in the event of mail server failure
- High-risk messages should only be dispatched over external networks by e-mail when
  protected by strong encryption
- Virus protection must be strictly applied and the techniques of protection reviewed regularly

12

Secure Virtual Private Networks

12.1 Scenario

The main problems for geographically distributed networks have been the need for a special infrastructure to handle remote access and the high cost of communication between nodes. A similar problem exists for users of a local network who require access when away from their place of work, such as travelling on business or working from home (see Figure 12.1).

When opening up a network for remote access, great care needs to be taken to ensure that only authorised users can gain access to the network, and all unauthorised users are locked out. This involves setting up a separate system of modem pools and telephone lines together with appropriate security measures such as firewalls, in addition to the standard system infrastructure for e-mail, fax and Internet facilities provided for internal network users. Despite falling telecommunication charges, the cumulative cost of leased lines and/or long distance and international calls for even small networks can amount to a staggering sum over a year.

The solution, which has only become available recently, is the use of virtual private networks (VPN), which uses the Internet [or other internet protocol (IP)-based network] as the communication means and a special secure protocol, which is transparent to the users.

12.2 Definition of VPN

Before we go any further, let us define what we mean by a virtual private network.

A dictionary defines 'virtual' as 'so in effect, though not in appearance or name'. So, for something to be a virtual private network, it should act like a private network yet not be one. Since all distributed networks are virtual to some extent, as they do not rely on hard wiring, what is meant by a private network? Once again, the dictionary defines private as 'secret, not public'. So, a private network is one that is dedicated to a group of users and is kept hidden from anyone else. Therefore, a virtual private network is one that acts as though it is dedicated to a group of users and is hidden from anyone else while operating over a public network. Since our public network is the Internet (or another IP-based network), which is notorious for being an attacker's playground, the important point is that the network is hidden to outsiders.

The concept behind VPNs is that a node can join the network for any desired function at any time and for any duration. The payload remains hidden by putting the data within an IP

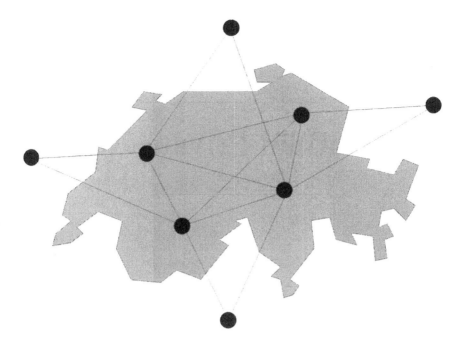

Figure 12.1   Network nodes

packet, encapsulating the entire packet, and putting this in a new IP packet. This new IP packet is addressed to the destination entry point and transmitted across the public network. The encapsulating information is stripped off upon arrival at the target entry point and the payload forwarded to its true address on the private network. This process is termed tunnelling and is illustrated in Figure 12.2.

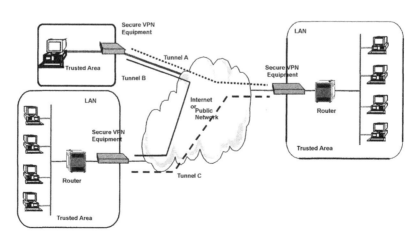

Figure 12.2   The process of tunnelling

12.3 Protocols

Two tunnelling protocols may be used: L2TP (layer 2 tunnelling protocol) and IP Sec (IP security protocol).

L2TP combines a number of existing technologies to create manageable dial-up networks. For the most part, L2TP does not claim to offer security. There are two proposals for gaining security: using IPSec in its transport mode or using a much weaker though, in some cases, adequate, point-to-point (PPP) security. L2TP, as its name implies, tunnels a link-layer protocol over IP. This allows for support of multiple protocols over an IP network, such as IPX or AppleTalk. The connection management protocol within L2TP lets the network administrator control the valid L2TP links. L2TP is targeted for remote clients, but some servers, routers and gateways will support it for network-to-network links. L2TP may not be common in firewall products, as its security is not recognized as fully secure.

IPSec provides network-level security for IP. Its management protocol, ISAKMP/Oakley, is also a security protocol and protects against man-in-the-middle attacks during the connection set-up. ISAKMP/Oakley defines the key agreement used for the encryption during set up of the tunnel. IPSec used in hosts as OS components or BITS ('bump in the stacks') implementations can work with gateway or router implementations, such as BITW ('bump in the wire') to create secured, dial-up network connections.

The distinction between L2TP and IPSec is important. L2TP supports dial-up network connections that can be secured. IPSec provides security that supports dial-up network connections.

Since we are talking about Secure VPN, i.e. using encryption, we will concentrate on IPSec. The following description does not take into account enhancements provided by the new IPSec RFCs (24xx). Despite adjustment to certain details, the overall philosophy remains unchanged.

RFC 1827, IP encapsulating security payload (ESP), describes two methods for using encryption to guarantee the integrity and confidentiality of data sent via the Internet (or via a private IP network):

These are 'Tunnel-mode' and 'Transport-mode'. The following definitions of 'Tunnel-mode' and 'Transport-mode' are from RFC 1827:

> In tunnel-mode ESP, the original IP datagram is placed in the encrypted portion of the Encapsulating Security Payload, and that entire ESP frame is placed within a datagram having unencrypted IP headers. The information in the unencrypted IP headers is used to route the secure datagram from origin to destination. An unencrypted IP Routing Header might (for IPv6) be included between the IP Header and the Encapsulating Security Payload.
>
> In transport-mode ESP, the ESP header is inserted into the IP datagram immediately prior to the transport-layer protocol header (e.g. TCP, UDP, or ICMP). In this mode, bandwidth is conserved because there are no encrypted IP headers or IP options.

The idea behind tunnels is that the entire packet from a node in the private network is encrypted and put into a new IP packet that has the address of a cipher system on another node in the private network (and, if needed, the way to route the packet to this system). When the packet arrives at the remote end of the 'tunnel', its contents are decrypted, and the original packet is sent on within the remote node's local (trusted) network. When tunnel-mode is used, hosts on trusted networks are hidden from hosts on the public network. Routing information need not be exchanged between the trusted networks and the public network.

In general, tunnel mode offers more flexibility and more security but is more complex to configure than the transport mode. The implementation of a single tunnel is shown in Figure 12.3.

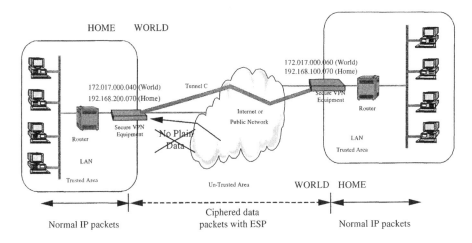

Figure 12.3   Implementation of a single tunnel for the example in Figure 12.2

All security implementations must be media-independent and fully compatible with IP routing (RFC 1812). When a packet arrives from the trusted network, IP examines the destination address and uses the private routing table to choose the tunnel cipher key needed to encrypt the data. It then gives the packet to the encryption subsystem. The public routing table is then used to route the packet through the Public Network.

12.4 Packet Header Formats

The following definition is from RFC 1827:

The Internet Assigned Numbers Authority has assigned Protocol Number 50 to ESP. The header immediately preceding an ESP header will always contain the value 50 in its next header (IPv6) or protocol (IPv4) field. ESP consists of an unencrypted header followed by encrypted data. The encrypted data include both the protected ESP header fields and the protected user data, either an entire IP datagram or an upper-layer protocol frame (e.g. TCP or UDP) (Figure 12.4).

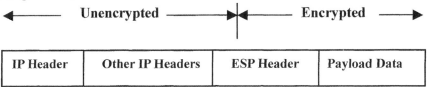

Figure 12.4   The encrypted IP datagram

The 'Other IP header' field is absent in IPv4 but may be present in IPv6. The opaque transform data in the ESP header is specific to the cipher algorithm used. Some or all of it may be encrypted. Note that some RFCs (e.g. 1827) refer to the security parameters index (SPI) as the 'security association identifier'. This is misleading. RFC 1825 states that a security association is defined by a destination IP address and the security parameters index.

Consider Figure 12.5; the SPI is a 32-bit value identifying the security association for this datagram. The value must not be zero. It provides a unique index into the security association list (SAL) discussed below.

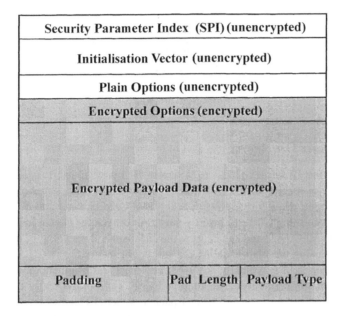

Figure 12.5   Detailed diagram of the ESP header

The size of the initialisation vector field according to RFC-1825 and RFC-1827 must be a multiple of 32 bits.

The plain and encrypted options may be absent.

The size of the payload data field is variable. Prior to encryption and after decryption, this field begins with the IP protocol/payload header specified in the payload type field. Note that in the case of IP-in-IP encapsulation (payload type 4), this will be another IP header.

The size of the padding field is variable and depends upon the size of the block cipher algorithm used. Prior to encryption, it is filled with unspecified implementation-dependent (preferably random) values, to align the pad length and payload type fields at an eight-octet (DES) or 16-octet (e.g. AES) boundary. After decryption, it must be ignored.

The pad length field indicates the size of the padding field. It does not include the pad length and payload type fields. The value typically ranges from 0 to 7 but may be up to 255 to permit the hiding of the actual data length.

The payload type field indicates the contents of the payload data field, using the IP protocol/payload value. Up-to-date values of the IP protocol/payload are specified in the most recent 'assigned numbers' (RFC-1700).

In tunnel mode, the complete IP version_4 packet format is shown in Figure 12.8. Figure 12.6 shows the structure of the unencrypted part and Figure 12.7 shows the structure of the encrypted part prior to encryption.

Version (4)	IHL (20)	Type of Service	Total Length	
Identification			Flags	Fragmentation Offset
Time to Live		Protocol (50)	Header Checksum	
Source Address				
Destination Address				
Security Parameters Index				
Initialisation Vector				

Figure 12.6   Unencrypted part of tunnel mode IPv4 packet format

The length of the IP header (IHL) is typically 20 bytes; most IP packets do not have options. The type of service field may be used by routers to make decisions about priority, delay and throughput, when forwarding a packet. KA9Q/NOS, for example, uses the type of service value to determine where to insert a packet in the output queue. Note that unless the router implements RFC 1827 and can decrypt the encapsulated packet, type of service routing can only be done using the plain data header and is not secure.

The identification field is typically not used. The flags and fragmentation offset fields are used to implement fragmentation. The time to live (TTL) field is really a hop count. It is decremented every time a router forwards a packet. When TTL goes to zero, the packet is discarded. This prevents a packet bouncing back and forth forever when a configuration error results in a routing loop.

Version (4)	IHL(20)	Type of Service	Total Length	
Identification			Flags	Fragmentation Offset
Time to Live		Protocol	Header Checksum	
Source Address				
Destination Address				
User Data				
padding (to align)			Pad Length	Payload Type (4)

Figure 12.7   Encrypted part of tunnel mode IPv4 packet format prior to encryption

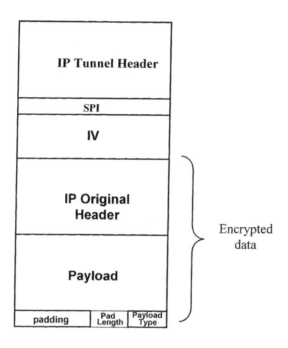

Figure 12.8   The complete ESP packet

When creating the tunnel header, the TTL field should be set to a user configurable value (e.g. 64). The TTL for the original packet should not be used, since it may reveal information about the trusted network topology.

Checksum calculations are discussed in RFCs 768, 791, and 793. In general, routers do not need to recalculate TCP or UDP checksums. For tunnel mode, there is no need to calculate a new checksum for the TCP header (the whole segment will be encrypted, including the checksum) or for the UDP header (the whole UDP datagram will be encrypted, including the checksum). When operating in tunnel mode, a new IP header is created, and a new header checksum is calculated.

In IP Version 6, the IP header will not have a checksum. Instead, hosts will be required by UDP, TCP and ICMP to calculate checksums over an IP pseudo-header and the entire remainder of the packet. Since the pseudo-header does not include the hop limit (the IPv6 name for TTL), routers typically need not recompute these checksums.

12.5 Security Association List

The security association list (SAL) stores the cryptological data needed to encrypt and decrypt IP packages, to mark each encrypted packet with an SPI and to time supervise the security association. In addition, variables of each entry are used to handle the replay protection mechanism and to indicate the use of compression. Each row of the SAL defines a security association. The table is initially empty, and all table entries are automatically created by the key agreement procedure.

The index to the table is the SPI_{RX}, the security parameter index for received datagrams.

This index is created by the receiving cipher system and is unique in the SAL. The key agreement transfers it to the sending cipher system, where it is entered in its SAL as the SPI$_{TX}$. For the sender, the combination of SPI$_{TX}$ and destination address (through the tunnel table) defines the security association. For the receiver, the SPI$_{RX}$ defines the security association.

Each entry is individually time supervised. When the timer expires, the entry is deleted or replaced by another.

12.6 Tunnel Table

The tunnel table (see Table 12.1) stores the destination address and SPI reference of the tunnel. Each row of the tunnel table is used to define one tunnel. The user must manually enter (via the MMI or over SNMP) the IP address of the secure VPN equipment at the other end of the tunnel. All other entries in the tunnel table will be made automatically.

Table 12.1 The tunnel table

Tunnel ID	Temporary	IP address tunnel end	SPI$_{RX}$	Tunnel status
1	False	192.5.5.3	107	Up
2	False	192.7.6.2	148	Down
3	True	192.123.77.4	23	Up
4	False	192.123.244.1	62	Up

Unknown IP addresses at the far end of the tunnel (for dial-in applications) must not be manually entered in the tunnel table. In this case, a temporary entry will be made automatically after the key agreement procedure.

At least one secure VPN equipment must have a tunnel table entry for the other side of the tunnel, or else the key protocol cannot be executed.

If a new tunnel table entry is made (manually or automatically), the new IP address will be added to the private routing table indicating the newly generated tunnel.

12.7 Routing Tables

Two routing tables have to be supported: a private routing table for one or more trusted networks, which may be hidden from public network, and a public routing table that is visible to the outside world. All routes from the trusted area are hidden from the public network (no entry in the public routing table), and only the own IP address is issued to the untrusted (public) side (one entry in the public routing table).

The public routing table is used to distribute the own IP address to the public network.

The private routing table is hidden from the untrusted (public) network side. The private routing table defines what IP addresses should use what virtual interface (tunnel). The private routing table can be updated manually or over a routing protocol.

The choice of supported protocols, both of which must be supported, are:

- OSPF (RFC1583).
- RIP-II (RFCs 1721-1724). The use of RIP-II for the public routing is configurable as a supplement to OSPF, i.e. OSPF remains the protocol of choice for maintenance of the private routing table.

Both the IP routing layer that switches IP packets and the OSPF routing protocol that updates the routing table should treat tunnels like output interfaces. When a packet is received from the trusted network, the secure VPN equipment looks in its private routing table and selects the appropriate tunnel. The output function for that virtual interface uses the security association to encrypt the packet and add the plain data header with the destination IP address of the other end of the tunnel. The public routing table is then used to route this plain data header and ciphered payload via the public network.

IP packages arriving from one tunnel to a secure VPN equipment must not be forwarded to another tunnel. This is in order not to override the security mechanisms, which selectively exclude access within portions of the VPN. If the IP destination address of the received packet does not belong to the secure VPN equipment, the packet is discarded.

Some nodes may be used in conjunction with a dial-up PPP connection to an internet service provider (ISP). In general, the ISP will assign an address to the node using PPP/IPCP address assignment. The node uses its logical location identity (LLID) and an authentication mechanism to automatically create a security association, associated with a tunnel that is needed to join a secure VPN.

'Normal' address resolution protocol (ARP) is required for basic operation of IP on an Ethernet. It is used to translate an IP address to an Ethernet address prior to sending a packet via Ethernet hardware.

A device that implements IP version 4 is required to act as a host (RFC 1122) or as a router (RFC 1812). A minimal conforming router is actually the simplest of these devices and is required to implement the IP layer, including ICMP and ARP.

The secure VPN equipment must also support proxy ARP (RFC 925 and 927). It must be able to use the routing table to automatically proxy ARP, on a per-interface basis, for a range of remote host addresses, so that hosts on two Ethernets separated by the secure VPN equipment appear to the IP layer as if they are on the same Ethernet. The secure VPN equipment must only provide Proxy ARP service towards networks on the trusted interface.

12.8 Packet Filtering

In a VPN, plain connections are generally not allowed. The user can define rules that, if fulfilled, cause a specific activity. This facility should support filtering on source and destination address, source and destination port, protocol type, and protocol flags. It should be able to filter input and output separately on a per-interface basis.

Example of rule definition:

1. Select IP source address xxx.yyy.zzz.nnn
2. Select IP destination address uuu.mmm.fff.eee
3. Select protocol type gg

   Activity: IF 1 AND 2 AND 3 THEN <bypass encryption>
   The rule compilation must support:

- Selection of specific IP address
- Selection of a set of IP addresses, i.e. masking of IP addresses
- Selection of specific protocol type
- Selection of specific port
- Selection of protocol flags, e.g. TCP SYN and ACK bits
- Selection of actions:

  - bypass encryption
  - discard packet

The advantage of filtering on protocol port, types, and bits is that the type of connection and whether a new connection is inbound or outbound can often be determined from this information.

The benefits of filtering on a per-interface basis and filtering on input and output are that it provides the ability to control access to the filtering devices itself. It simplifies filter administration. For example, all filters can be applied to the interface connected to the public network. Filtering on input allows the device to efficiently discard packets sent as part of a penetration attempt or a denial of service attack. Filtering on output helps prevent information from leaking to the public network.

When filtering on protocol ports and bits, IP fragments present a special problem. The first fragment contains the protocol header, but other fragments do not. One solution to this is to maintain a list of recent IP source-destination address pairs and protocol type and cache the filter decision made on the first fragment. When filtering on some protocol information in addition to the IP header, such as source or destination port number, the secure VPN equipment should be able to cache the decision made on the first fragment and apply that decision to subsequent fragments that have matching source and destination address and the same protocol type. The behaviour should be one of three options that are configurable on per-interface basis. The options are: (1) discard all fragments, (2) apply decision from cache, and (3) accept all fragments. The default behaviour must be to discard all fragments when doing protocol-specific filtering for an interface.

The filtering rules can be applied on different points in the processing of an IP package, and it must be possible to define separate rules for the following four points of filtering.

- Untrusted-encrypted filter. After reception of the package from the untrusted interface, but before decryption and routing.
- Untrusted-decrypted filter. After decryption and removal of the tunnel header of an IP package from the untrusted interface and after routing. The filter is applied to the encapsulated IP packet.
- Trusted-decrypted filter. After reception of the package from the trusted interface, but before encryption and routing.
- Trusted-encrypted filter. After reception of the package from the trusted interface and after encryption and routing.

12.9 Threats and Countermeasures

The threats to a secure VPN system can be divided into three areas:

- Attacks within the public network
- Attacks within the nodes of the trusted network
- Attacks from the public network on the entry points to the nodes of the trusted network

12.9.1 Attacks Within the Public Network

Once the data leave the trusted node of the private network and until they arrive at the destination node's entry point, they are exposed to attacks from anyone who also has access to the public network. It is within this area that the data can be intercepted, viewed, modified, or recorded and retransmitted.

The secure VPN system deals with these threats by:

1. Encapsulation and encryption of the data including the original header.
2. Manipulation and replay detection built in to the used protocol.

Accordingly, no information can be gained from intercepted packets, and any manipulated or replayed packets will be discarded by the receiving equipment.

12.9.2 Attacks Within Nodes of the Trusted Network

Attacks within this area are common to all local networks and have been discussed elsewhere in this book. The organisation and implementation of security within this area are the responsibility of the local network and security managers and must be assumed to be a trusted area. The secure VPN equipment will only allow authorised hosts within the local network to access remote nodes via tunnels.

12.9.3 Attacks Aimed at Gaining Access to the Private Network

Since the secure VPN equipment presents its own (and only its own) IP address to the public network, attacks may be directed to the equipment itself with a view to gaining access to the local network behind it or to flood the equipment with packets, thereby denying access time to authorised users.

Physical attacks on the secure VPN equipment form part of the security for the local network as described above. Logical attacks from the public network are handled by the filtering of packets performed by the secure VPN equipment. It is therefore necessary that the rules for filtering fully address these threats. In particular, a means to prevent denial of service by repeatedly sending legitimate packets, such as pings (ICMP echo requests), should be included.

12.10 Example Application – 'Diplomatic Network'

A good example of an application using secure VPN would be a diplomatic network in which all its missions throughout the world communicate with its Ministry of Foreign Affairs using e-mail. In addition, when diplomats travel to various summits, etc., they can stay in touch by dialling in to the network as required.

For the purposes of this example, we will define four types of node:

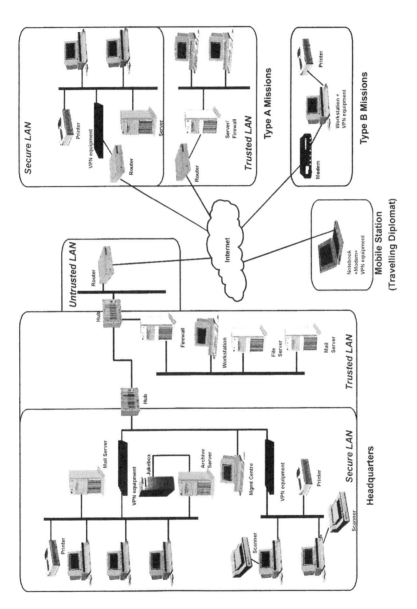

Figure 12.9 'Diplomatic' network

1. Headquarters: representing the Ministry of Foreign Affairs.
2. Type A mission: representing major regional missions with their own local network.
3. Type B mission: representing minor missions operating from a single host.
4. Travelling diplomats: representing individuals away from their normal place of work.

The 'post office' for all users of the e-mail system is held on the mail server at the headquarters and therefore acts as the central node in a star network. Communication between nodes is therefore performed using tunnels to and from the headquarters.

The headquarters and the type A missions have a permanent Internet connection (leased line), while type B missions and travelling diplomats use dial-up connections via a local ISP.

Figure 12.9 provides an overview of the 'diplomatic network'.

At the headquarters, the LAN is logically divided into three different networks:

Untrusted LAN: containing the router for Internet and PSTN access plus any public Internet surfing stations and public Web Server

Trusted LAN: containing existing file and mail servers for workstations handling nonsensitive data. It is protected from the untrusted LAN by the firewall

Secure LAN: containing servers and protected workstations for handling sensitive data. It bypasses the trusted LAN and is protected from the untrusted LAN by VPN equipment.

For missions that do not contain a LAN, workstations that handle sensitive data are individually protected by built-in VPN equipment. Connection to the network for these workstations is provided by dialup access to the Internet.

All missions are provided with disk encryption facilities, anti-virus software and backup facilities.

Network and security management of the VPN equipment is performed by the management centre at the headquarters. The management centre is able to control and monitor all units in the network that are online thereby providing a continuous overview of the entire communications network.

At the headquarters all messages sent or received by secure LAN workstations are automatically copied to the archive system for processing and storage in a database held on the optical jukebox. The archive system is also equipped with scanners for handling of paper documents.

Within the system, confidential data are always protected. They are stored encrypted on the secure PCs, encrypted by the secure e-mail system, and further encrypted when communicated via the Secure VPN equipment.

Key management for such a system is a major task, so to reduce the number of keys involved, the e-mail system uses node-specific keys rather than individual user keys. These keys are linked to defined addresses in each user's address book for automatic selection when creating messages. This means that users do not have to worry about which key is required for which address. This still means that there are a lot of keys, thus, with periodic replacements, presenting a problem for the archiving system. This problem is overcome by the archive workstations deciphering the received copies using the current key and reciphering them using a special 'archive' key before storage in the archive database. This means that messages can be retrieved from the database even long after the key that was originally used to cipher the message has been removed from the system.

13

Military Data Communications

Chapter 3 covered military voice communications security in detail, and some mention was made there about data operations. There is quite a deal of common ground as voice and data terminals serve within the same military, field environment and the modes of transmission are usually the same. However, there are enough technical and operational differences to warrant a short chapter on field data communications, especially as there is such an influx of new military standard data terminals on the market. The features they carry offer a sophisticated support for the traditional voice packages, and much interest is generated by the data, graphics, GPS, packet data comms and Internet services that the new technology brings to the battlefield.

Dr David Callaghan of Thales Defence Communications captures the classic approach to cryptographic design philosophy, which is particularly significant in military battlefield communications, but certainly not specific to the scenario.

Cryptographic design philosophy:

> Thales Defence Communications' (formerly Racal) design philosophy for military and parami-litary radio equipment stems from a good understanding of the communications environment and the need to design for the worst case threat scenario.
>
> Military and paramilitary equipment are often characterised by narrowband half-duplex communication channels in a tactical deployment. This leads us to design symmetric key cryptographic systems, which do not need numerous message exchanges to set up a communication, with very robust synchronisation schemes to withstand the hostile nature of the military tactical environment. The philosophy is to ensure the synchronisation systems are fully automatic in order to guarantee ease of use under all operational circumstances.
>
> In order to overcome the difficulties of physical key distribution it is now common for military and paramilitary equipment to have built-in OTAR (over the air re-keying) capability, even though it is recognised this may not be the normal method of distribution for many users.
>
> The design of the cryptography embedded within Thales Defence products assumes a worst case scenario where an enemy has knowledge of the algorithm and some recordings of plain and cipher text. The cryptography is designed to have good local and global randomness properties and non-linearity such that the only method of attack is a brute force key search.
>
> In addition to proprietary algorithm, design philosophy makes allowance for the customisation of algorithms to provide unique solutions for different user communities. This is achieved by customising the non-linear parts of algorithms to ensure that output keystream is unique even if all keys were to be used...

13.1 Applications

As battlefield technology increases, the need for fast and reliable data transfer becomes even more critical, even at the lowest levels of an organisation. The use of data provides for a better utilization of the transmission media and the different applications often require differing services and different methods of protection. These services can be generally identified as shown in Figure 13.1.

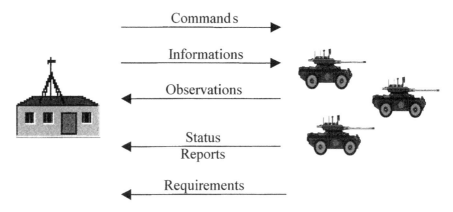

Figure 13.1   Battlefield communication services

13.1.1 Data Over Radio Links

A common requirement of military communications is the ability to transmit beyond line of sight, and where telephone wired circuits are not available, high-frequency (1.5–30 MHz) (HF) radio is very often the choice medium. HF radio relies upon its propagation by the refraction of its signals by the ionosphere. The quality of transmission is often disturbed by fading, distortion, low signal-to-noise ratios and interference from other channels. To cope with these problems, a high-quality modem is required to combat data loss over the ether, which can be a major disruption to message transmission.

Compared with HF, very high frequency (30–300 MHz) (VHF) and ultra-high frequency (300–3000 MHz) (UHF) frequencies enjoy frequency modulation and line-of-sight transmission, which makes for a much more stable transmission platform. A useful feature of a radio modem operating in a military environment is the ability to compensate automatically for a tuning differential between two transceivers. This can often occur where older radio transceivers are employed. An automatic adjustment within ± 75 Hz is a useful tool to have available.

13.1.2 Modes of Radio Operation, Automatic Repeat Request and Forward Error Correction

The two principal modes of radio operation are automatic repeat request (ARQ) and robust-forward error correction (R-FEC). For communications between two transceivers, ARQ provides for an error-free data transmission by using a handshake between the two stations

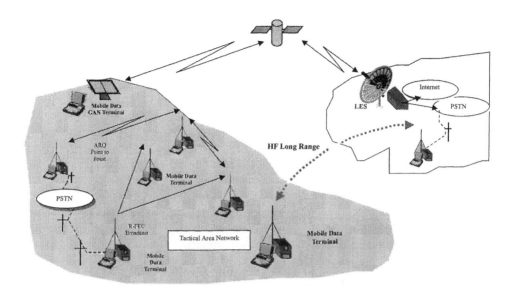

Figure 13.2   Data over radio links

to accommodate any discrepancy between the transmitted signal and the received signal. Errors detected by the receiver will bring about a retransmission of the offending data package, and a high repetition rate will prolong the transmit time. However, this duration can be substantially reduced by using a modem, which optimises its modulation technique according to the propagation characteristics. Such a modem can represent a big saving in transmission time.

As ARQ relies upon a handshake between two stations to correct errors, it cannot be used for broadcast transmission to a number of stations, and in this case, R-FEC provides the solution. The FEC breaks the data signal into 3-s-long packets, which it retransmits a number of times varying between 0 and 7, according to the propagation climate and associated programming. With FEC, the repetition transmissions take place irrespective of whether or not the receiving stations actually receive the packets, intact. The transmitting station does not receive any acknowledgement that its transmission was ever received. Naturally, the retransmission incurs time penalties, but what it loses in time, it gains in transmission robustness. Another advantage, and indeed a necessity, of FEC is that the receiving station does not respond to the reception of a message as ARQ requires, and this factor makes FEC most useful in situations where radio silence is essential.

For efficient and reliable transmission of data, the terminal must provide for the setting up of the transmission medium by providing the means to program transmission parameters as required by the type of radio and propagation.

13.1.3 Use of GAN Terminals in Battlefield Applications

Recently, global area network (GAN) satellite terminals have made inroads into the military scene, and their flexibility, portability and quality of service provide an exciting alternative to the traditional battlefield radio environment. The subject of satellite communications is dealt with in greater depth in Chapter 4, but Figure 13.3 shows a typical army application with a 'notepad terminal'.

Figure 13.3   Army data transmission over GAN satellite terminals (courtesy of Nera Telecommunications Ltd)

13.2 Data Terminals and Their Operating Features

The hardware varies from the simple hand-held data terminal to a 'ruggedized' laptop. The former typically comprises an alphanumeric keyboard, an LCD screen displaying two to five lines of text and a modem or interface unit that allows connection to a radio transceiver.

The ruggedized laptop terminal opens up numerous possibilities with all that operating systems such as Windows can impart. One would expect to find a software package including:

- Word processing
- Graphics
- Data
- Imaging
- Internet communicating
- Spreadsheet
- Video
- Voice

- Security programs for local file encryption and ciphered telecommunications.
- File management

Software packages enable e-mail to be sent over tactical VHF radio networks using standard packages such as Windows 95™ software running Microsoft Exchange, Windows Messaging or Outlook (see Figure 13.4). The ability to transmit maps, pictures, text, video and identification material, by e-mail, etc., coupled with full compatibility with desktop computers, signals the emergence of a different communications era, as far as battlefield data are concerned.

The use of these laptop field terminals allows the cryptographic tools developed for PC security to be applied in this environment, and Chapter 10 covers this in some depth.

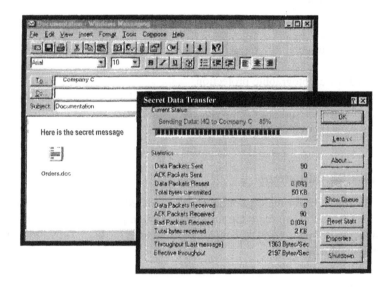

Figure 13.4   Windows-based programs, ideal for data messaging

13.3 Technical Parameters

Hardware components of a 'notepad'-style terminal should consist of:

- Ruggedized hardware platform
- Microprocessor
- Hard disk
- Efficient RAM capacity
- Floppy disk drive for storage and import/export of files
- Input interfaces for peripherals, e.g. printer/scanner
- Communication interfaces:

  - Telephone
  - HF/VHF/UHF radio compatibility
  - GAN satellite terminals

- GSM
- X25, LANs, etc.

- Robust, built-in modems

  - Point-to-point and broadcast functions
  - Automatic error correction
  - Automatic compensation of TX/RX frequency offset
  - Long-range HF communications

In either case, the prime considerations of the customer are:

- End-to-end communication security
- High operational security
- Efficient carriage of strategic $<->$ tactical data
- Robustness of logical links
- Environmentally suitable, i.e. ruggedized, portable and constructed to MIL Standards 461C and 810E
- High standards of cryptographic protection
- High degree of compatibility with existing systems and ex-battlefield environments
- Comprehensive and flexible key management operation
- Removable security modules

13.4 Security Management

The security risks to the information stored on or transmitted from a data terminal are sixfold, i.e. the unauthorized access to data, the unauthorized manipulation and use of data, loss of data due to deletion or theft, physical damage to the storage medium, TEMPEST monitoring and cryptographic attack.

13.4.1 Access Control

In a battlefield scenario, there is a high risk of hardware falling into the hands of the enemy. This is certainly also true for communications equipment, including mobile data terminals. Wherever such circumstances are likely, special precautions must be taken to remove the possibility of the enemy gaining information from a data terminal. In addition to the use of passwords and access tokens as described elsewhere in this book, users should be advised to adopt the following practices:

- Terminate active sessions immediately after they have been finished with
- Log off when the session has finished and do not simply rely on powering off the data workstation
- Use screen savers with password protection
- Be familiar with emergency clear functions including 'wipe' or 'burn' programs to destroy 'erased information'
- If the data terminal has to be abandoned, remove any security module, e.g. the PCMCIA card, and if its seizure by the enemy cannot be prevented, ensure that it is destroyed.

13.4.2 Data Encryption

There are two aspects regarding data encryption to resist cryptographic attack. They are the encryption of the files stored on the storage medium and the other, the secure transmission of files. Both threats can be countered by the encryption of files stored on the terminal disk, and this can be achieved as illustrated by Figure 13.2. The encryption uses a one-time secret key and a CMK master key.

13.4.3 Loss of Data

The loss of data due to deletion, intentional or otherwise is controlled mostly by the access procedures mentioned above. The loss of data due to the theft of a terminal on the battlefield poses a different problem. Unless copies of the lost data files have been transmitted to another station, they will be irretrievable. However, the main fear is that the data will become available to the foe, and so, the user has to rely upon the strength of the encryption and the access controls that have been implemented with just such a situation in mind. The same fears and solutions are applied in the case of the terminal or, more specifically, the storage medium, being physically damaged. One would expect military data terminals to be manufactured with robustness and durability as a basic design strategy, but logical security should not rely on this.

13.4.4 TEMPEST

Protection against the compromising radiation of sensitive data due to electro-magnetic compatability (EMC) or TEMPEST problems is dealt with in Chapter 1, Section 1.5.

13.5 Key Management

A number of useful key features should normally be available to the key manager, as illustrated in Figure 13.5. A Windows-based management tool would include a display illustrating the generation of a customer master key (CMK). This key is often link-specific but may be group-specific and is used as a source key for the generation of session keys for that link or group. Normally, one would expect such a key to be 32 decimal or hex characters, giving a key length of 128 bits.

There would be an option here to either enter the key manually or use the management program's internal random generator, after which the generated key data could be accepted or thrown out as preferred. A key signature should be produced for each set of key data generated, and these can be used to verify that common keys exist at each end of the communication link. In addition to the signature, each key might well have a label or ID and an extension that are both useful in actually identifying the purpose, i.e. a channel key, and may also be used as a component in the formulation of the key signature. The key validity may also be part of the extension and, if so, may be readily available to aid key management purposes. Ideally, a pool of master keys will be generated for the formulation of a network hierarchy and would be displayed to the key manager for them to assign as required. Here, a list of all key parameters can be listed along with facilities to add or delete individual master keys as desired. It would also be interesting to see an

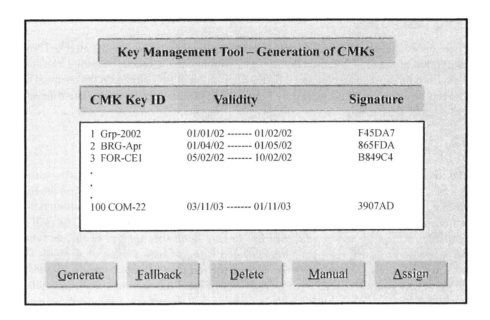

Figure 13.5   Typical screen display of a key management program

option to use a fallback key be enabled in case of an emergency situation arising. A function allowing the deletion of all keys is useful as an emergency clear. Key selection by a user may be either automatic by linking the message to a particular destination or manual, as preferred.

13.6 Combat Packet Data Networks

Communication of battlefield data has traditionally presented significant challenges, with every sensor, weapon or command and control system requiring a dedicated frequency allocation, a dedicated radio bearer and a dedicated interface. The Thales Panther family of enhanced digital radios (EDRs) makes optimum use of the frequency spectrum, by sharing the communications channel efficiently among a number of independent data users.

13.6.1 Packet Radios

The Panther EDR family is a multi-media combat net radio (CNR), which meets the challenges presented by the pace of change in battlefield digitisation. Two recent significant extensions of Panther EDR's capability have increased its strong position on the digital battlefield, thereby maintaining its value to commanders in the field. The first extension was the provision of a fully distributed packet data facility. The second was an embedded GPS receiver and a range of services that exploit GPS information and can be used with embedded or external GPS. The basic concept of packet radio is illustrated in Figure 13.6.

Regions of Radio Connectivity

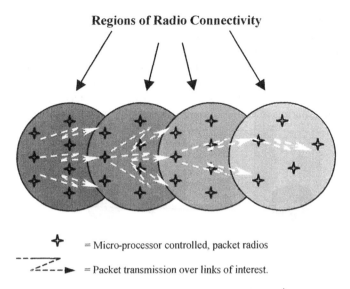

✛   = Micro-processor controlled, packet radios

- - -➤   = Packet transmission over links of interest.

Figure 13.6   The basic concepts of packet radio (courtesy of Thales Defence Communications)

13.6.2 Packet Data Networks

The Panther EDR has an extensive range of resilient data facilities backed by application level software for their enhancement. These include multiple simultaneous voice and data calls utilising Panther EDR's multiple simultaneous access (MSA) and carrier sense multiple access (CSMA) features. In addition, the EDR provides a package short message service (PSMS), which facilitates seamless transfer of short, pre-formatted messages between radios. Packet techniques are already used at the application level to provide a guaranteed service to users when sending large files or imagery across a radio link

It is worth recalling the reasons why packet radio techniques are of interest to military users. Battlefield data have traditionally presented significant problems with every sensor, weapon or command and control system requiring a dedicated frequency allocation, a dedicated radio bearer and a dedicated interface. Spectrum allocation is a serious concern because military users have to compete for spectrum space with commercial and telecommunications enterprises. They also require communication links that are resistant to electronic warfare and, in turn, demand more spectrum. Efficient exploitation of the allocated spectrum is thus of paramount importance.

The dedicated radio bearer and interface represents a cost and space penalty and total lack of flexibility. It also leaves users 'locked in' to special-to-type legacy systems and standards, totally unable to take advantage of advanced technology and performance as these become available.

Packet radio systems make better use of the spectrum by sharing the communications channel efficiently among a number of independent data users. This sharing is made possible because of the typical variable data transmission requirements of many users who do not require continuous 'data pipes'. Effective packet radio provides spectrum-efficient, adaptable, affordable and extendable solutions to a number of battlefield data transmission requirements.

A successful packet radio network has a fully distributed architecture, with minimal management overheads. This allows each radio node to learn automatically about the network topology, users connected and route quality and availability. Each radio node can then make the best possible decision when determining the action to take on receipt of data packets off-air, sourced either locally, or via an internet connection (see Figure 13.7). This strategy minimises clashes and re-transmissions, leading to a better network capacity and better sharing of traffic among the available routes between data sources and destinations.

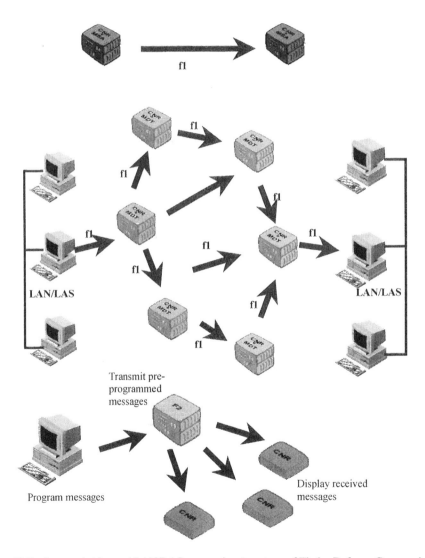

Figure 13.7   Internet bridge and LAN/LAS connection (courtesy of Thales Defence Communications)

The Panther EDR incorporates an intelligent data modem, optimised for use in hostile RF environments. The modem provides forward error correction, robust synchronisation and carrier sense multiple access (CSMA) for efficient use in a packet data network supporting both voice and data communications.

Panther EDR provides a MIL-STD-188-220 (B) compliant interface at the physical layer. The interface provides both non-return to zero (NRZ) and X21 interfaces between DCE and DTE, CSMA network access and network busy sensing and receive status indication.

Panther's network data service (NDS), provides intranet connectivity across a CNR net and Internet connection to local area networks (LANs). The service utilizes commercial off the shelf protocols such as TCP/IP to provide a transport mechanism for application connectivity. The architecture of the NDS enables a packet data application to be run on a PC-based message data terminal (MDT), which connects directly to the radio. The MDT can also act as a bridge between LANs or as a gateway to the Internet via a LAN (see Figure 13.7)

The network and link layer protocols for packet data routing and networking are currently available for fixed land-based networks, which do not require the issues of connectivity and mobility to be addressed. Network protocols for CNR are currently evolving, namely MIL-STD-188-220, and further developments will eventually yield improvements specifically addressing the mobility and connectivity issues evident in a mobile tactical network.

The architecture of the Panther NDS provides for the implementation of these protocols at the appropriate level in the protocol stack as and when they become available. The current implementation includes a protocol stack, which provides intranet routing across a CNR net with a 'store and forward' capability, achieved by the forward-looking software architecture and the use of open standards, both hallmarks of Panther EDR.

With a reliable battlefield medium assured, the actual data protection can be achieved by the encryption techniques, key management and access controls described in detail in Chapters 10–12 and elsewhere.

14

Management, Support and Training

In April 1943, things were not going well for the Japanese in the battle for the Pacific, against the American war machine. The tide had turned, and after the loss of Guadalcanal, the Solomon Islands were threatened as the USA built up momentum in its drive to the West. The Commander in Chief of the Japanese combined fleet, Admiral Isoroku Yamamoto, decided that a moral-boosting tour of inspection should be made to his troops of the region. Naturally, the visit of a Commanding Officer to any military base is a big event, and none more so when that commander was the national icon of the Japanese warrior, the revered Yamamoto. This man was the dominant figure within the Japanese Navy, a strong and imaginative leader and a real thorn in the American Navy's side. He was very much the perpetrator of Roosevelt's 'day of infamy', Pearl Harbor. What ensued was a classic story of cryptanalysis, the consequences of which were to have an indelible impression on the future development of the Pacific War.

David Khan, in 'The Codebreakers', writes:

At 17:55 on the 13th of April 1943, the proposed visit of Yamamoto to his front line troops was broadcast to the commanders of those bases intended as 'stop-offs' in a 'whistle stop', five-day tour. Naturally the broadcast of such an event related many details of the commander's itinerary, providing dates, times and addresses and emphasized the need to safeguard the great leader during his trip. It seems that such classified information was encoded in the high security code JN25, a well used and familiar Japanese Naval code of the time. Unfortunately for the Japanese, the allied codebreakers had made deep inroads into the code and whilst the analysis did not reveal all factors, the American officers recovered many details including the codes of bases such as RR for Rabaul, RXZ for Ballale, a small island in the Solomons group and RXP for Buin, a base at the southern tip of Bougainville.

The translated message (abridged) read:

The CinC, Combined Fleet will inspect the bases Ballale, Shortland and Buin in accordance with the following.
06:00 depart Rabaul on board medium attack plane (escorted by 6 fighters) to arrive Ballale at 08:00. Immediately depart for Shortland by subchaser arriving at 08:40, departing Shortland 09:45 by subchaser arriving Ballale at 10:30.
11:00 depart Ballale on medium attack plane, arriving Buin at 11:10, lunch at First Base HQ (Senior Staff to be present).

14:00 depart Buin aboard medium attack plane to arrive in Rabaul at 15:40. In the event of inclement weather, the tour will be postponed by one day.

The cryptanalyzed transcript was the death warrant for the highest Japanese commander. This was the closest that Yamamoto had ever come to the combat front and the Ballale-Shortland-Buin segment of the visit was within striking distance of the forces of Admiral William F Halsey. Halsey was given the task of destroying the Japanese force carrying Yamamoto.

The British and American forces were exceedingly careful about the use of information gleaned from the cryptanalysis of enemy messages and to protect their never ending source of 'war winning material', a cover story was usually invented. The obvious fear was that such an evidently pre-determined, pinpoint attack could only have come from the analysis of the Japanese radio transmissions. The debate was focussed on the JN25 code, which was classified by the Japanese, as being of high security yet was being almost daily read by the Americans. The introduction of a new code as a result of attributing the death of Yamamoto to communications cryptanalysis might have severe consequences for the future American success in the conflict. This was a particularly difficult decision to make, as they were just in the ascendancy. Having said that, the Americans, based upon recent past events were still confident that even if a new code was introduced as a replacement to the JN25, their analysts would be able to break into it in reasonable time. It says little for the Japanese effort at securing their communications and even less for their naivety. With such evidence at hand, the order was given to proceed with the intervention. As Khan writes 'the death warrant was now signed, sealed and delivered'.

To minimise the possible Japanese response of changing codes and tightening communications security, the Americans played upon a story that Australian observers had made reports of the Yamamoto flight after a visual sighting.

On the morning of the 18th April, 18 P-38 Lightnings took off to intercept the Yamamoto flight and right on time, over Bougainville, the American fighters discovered and fell upon the Japanese squadron. They closed within 3 km of the Japanese before they, themselves were sighted and attacked with single-minded determination. Yamamoto's plane was hit, lost a wing as its engine exploded and crashed into the jungle, taking with it, the life of Japan's favourite warrior. It was a severe body blow to that proud nation and his death had a distinct effect on the morale of the people. All but one of the Lightnings returned safely to base. Cryptanalysis had given the Americans a famous victory and for the Japanese another brick fell in its demolition.

The lesson to be learnt from this history is that when one communicates, one broadcasts to the world and beyond. Unless the information is protected by the armour plating of high security encryption and the communication practices of every individual involved in securing that message are both laid down and acted upon as fundamental laws, the message will be public. Those who are naive enough, as the Japanese were in the case of Yamamoto, to believe that nobody cares to, or is able to, listen should beware the consequences.

When this book was first envisaged, the author's proposal drew a comment from one of the team of reviewers, who did not see the necessity of this final chapter. The author assumes that the reflection came from an academic, and with all due respect, perhaps one with little hands on experience at the operational level. From his years of experience at the 'sharp end' of secure communications, the author has seen much evidence that the fine ideals from the cryptographer become clouded as they filter through to the man with his finger on the button. History has shown that a lack of attention to management procedures has resulted in the failure of even the best-conceived plans, and this fact underlines the necessity of this chapter.

Just as Yamamoto, many others have trusted those beneath to follow the procedures developed by higher authorities. The failure to ensure that those responsible for packaging and posting the nation's secrets did not so in the manner in which they were intended meant that they paid the price. This, then, is the realm and purpose of the security manager. The reader should judge for themselves if the preceding chapters present sufficient information and advice for their communications to be secured. It is the author's strong opinion that they are not and that the story is incomplete without some attention to the body or team who must pull all of the issues together. A management figure is required to develop and implement a practical strategy that has all the characteristics of integrity, from the highest echelon to the fellow who presses the button and so dispatches the secret into oblivion. The remainder of this chapter is dedicated to influencing not only that figure that goes by the title of 'security manager' but also to those who appoint them and those who should follow their direction. For this onerous task, there are management tools available to help.

14.1 Environments of Security Management

The setting in which the security manager and their organisation find themselves can actually be defined as two environments, the global environment and the task or local environment. Each has a distinct effect on the shape of his or her role.

14.1.1 The Global Environment

Consider Figure 14.1.

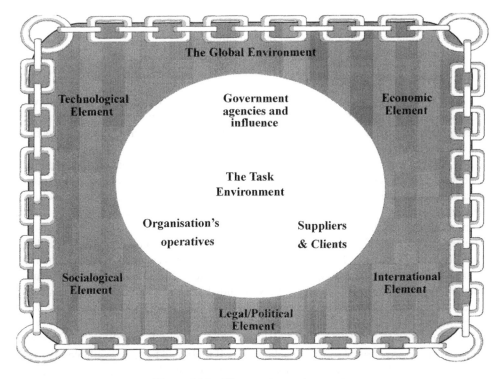

Figure 14.1 The security environments

14.1.1.1 The Technical Element

This book has been very much focused upon the technical element of the security environment. The author has, from time to time, emphasised the ways in which advances in technology affect both the cryptographer and the analyst. The introduction of faster and more efficient tools, particularly the increase in computing power, triggers new or more refined approaches to maintaining security.

The advances in electronics to produce innovations such as the ASIC and vocoder have had a marked effect on the security of voice encryption devices, as have the spread of mobile telephones and the emergence of global area networks supported by Inmarsat. Smart cards are relatively new to the security industry, and one sees their continuing technical progression having an increasing effect on key management. Biometrics and the like add further possibilities to manage systems. It is not only in the hardware technologies that new innovations rise to confront the security professional. Also in the field of cryptography, new algorithms and/or their applications have been evolving, largely as a result of the public debate, and sometimes not so public a debate, to generate ideas and concerns of those who wish to avail themselves of their security, or otherwise. What effect will AES have on the industry? It therefore transpires that the budding security manager must have and maintain a high interest in how the technical elements of the environment are going to affect their domain. Like Dönitz, the German U-boat commander, wishing to keep one step ahead of the bustling, analytical colony of Bletchley Park, England in the 1940s, saw the possibility and fruits of introducing a fourth rotor into the Enigma Machine. Conversely, the opposition, keeping their finger on the technological pulse to combat the increased difficulties of breaking the new machine configuration, developed Colossus, the first electronic, programmable computer. This element on its own demands a great deal of attention.

14.1.1.2 The Economic Element

If one considers the technical element as being a dominating factor in the environment, what is the consequence of finance? There are two aspects of the economic element, one more akin to the local environment and one aligned to the global environment. They are, however, closely interrelated.

The global aspect relates to the flow of funds through the international markets and exchequers. The currency crisis in Southeast Asia in 1997 brought about a collapse in trade and investment that affected organisations around the world. As with most technologies at that time, security manufacturers suffered the consequence of the downturn as clients withdrew from the security market.

The periods of regional or global recession have a knock-on effect with regard to the local aspect. There is obviously little confidence and motivation in investing in the development of an organisation's assets and subsequently the drive to protect them. This means that a security manager might be in a position where they do not have sufficient resources to achieve adequate protection of their networks. Financial cutbacks could mean a loss of staff and the inability to carry our technical maintenance or improvement.

Just how much worth is there in the data that the organisation has to protect? If it costs a company more in providing a security package than if their data became available to all and sundry, then why bother? Not to forget, of course, how much did it cost to generate those data in the first place? However, if the loss of confidentiality to a competitor results in the demise of the organisation, one has to question whether or not there is a limit to the efforts that should be made to preserve it.

Experience reveals that despite the most conservative estimates of security budgets, the security manager, forgoing strong arguments to the contrary, will be left with something less than ideal, to implement their network. Securing the organisation within the financial constraints is perhaps the biggest headache of them all.

14.1.1.3 The International Element

The international element includes the developments outside the organisation's home territory but has the potential to critically influence the organisation in question. The election of an influential national leader can instil both fear and confidence that promote the efforts of a region to either invest heavily in security or not. Such occasions where a threat to an organisation's assets is increased usually result in a sudden interest in security and subsequently a flow of funds to react to the threat. Conversely, the prospect of 'peace and goodwill between men' always has a detrimental effect on the defence security industry. However, commerce thrives, and the technology of security agencies and manufacturers shifts to explore a different niche.

14.1.1.4 The Legal/Political Element

Many consider that the communications security industry is fundamentally influenced by the control that the NSA has over the encryption products that are imported, exported or restricted, as far as the US are concerned. This control has meant that high-grade equipment manufactured in the US has not been available for export, and conversely, the import of the same products has been severely restricted. This has meant that American manufacturers have missed out on vast export sales, as only weaker algorithms such as DES have been available for foreign countries or organisations. For many, this has not been an acceptable solution to their security problems; indeed, the restriction on American algorithm export has raised, rather than eased, concern for prospective customers. Faced with a choice of buying weaker encryption and all that that entails, or searching the market for proprietary algorithms or custom-designed merchandise, many discerning consumers have looked elsewhere. As a result, a niche market in encryption algorithms and equipment prospers. Recent changes in the US's export restrictions have had a slight effect on the industry but very little as far as the foreign manufacturers of ciphering equipment are concerned. The recent revision (January 2000) of regulations (abridged) on encryption items and the Wassenaar Agreement includes the advice that there are no longer controls on:

- Goods using symmetrical algorithms with a key length of less than 56 bits
- Goods using asymmetrical algorithms RSA® or Diffie–Hellman with a key length less than or equal to 512 bits
- Goods for the mass market if all of the following points are fulfilled:

  1. Free, over the counter, like cash selling and Internet purchase, etc.
  2. Using symmetric algorithms with key lengths of 64 bits or under

3. Cryptographic functions not easily modified by the user
4. Developed for installation without any support

Export controls always exist for:

- All defence goods
- All ECCM/EPM (electronic counter counter measures) goods
- All compromising emanation protected goods

All of which make for interesting reading for both security managers and their suppliers.

Other major influences are changing political scenarios around the world. The Iraqi invasion of Kuwait and the Iranian revolution all had stimulating effects on the communications industry as individual nations sought to meet the challenges brought about by these events.

Once again, one must keep one's finger on the pulse and inform oneself about how the legal and political element may influence a network and the equipment with which it should be protected.

14.1.1.5 Sociological Element

Almost certainly, the greatest sociological change over the last decade, as far as the general public is concerned, has been the use of the Internet for shopping and trade, and the use of services such as Internet banking. It is not only the general public that has embraced the revolution, but commercial institutions, the military and even political procedures have been moulded around the new medium. This new method for doing business has opened up a whole new industry to manufacturer communication packages to meet the demands of the users. Naturally, as the technology evolved, so did the opportunities for exploitation and the naïve security manager, and there are many who would have been caught cold by the skilful exponents of corruption who dared to question the security of their networks.

14.1.2 The Local/Task Environment

In contrast with the global environment, over which the security manager has little, if any, influence whatsoever, the local or task environment should be very much under their control.

14.1.2.1 Suppliers

Despite the security export constraints of nations like the US, France and Britain, it is hopeful that both a diverse selection of equipment and algorithms will be available for consideration when designing a secure network. It is, of course, in the best interest of the manager to be in a position of strength when negotiating the supply of equipment and support. Therefore, the manager must be aware of, and investigate, alternative solutions and sources of security packages. The more options to hand, the stronger will be the manager's position in carrying out negotiations.

14.1.2.2 Clients

The security administrator of a larger organisation is likely to be influenced by different groups of people, each having their own ideals and requirements, and it is rare for a security administrator to be able to satisfy all quarters with general solutions. Failure to consider these intricacies will lead to dissatisfaction and will introduce tendencies for individuals and parties to go their own way and therefore undermine the authority of a central controlling body. The latter must be considered essential to high-security implementation.

14.1.2.3 Government Agencies and Influence

A security administrator, including that of private companies, cannot expect to work in isolation from the influence of the resident government and their security agencies. Governments have a strong desire to oversee most matters of communications security. From time to time, they exert pressure to impose their own policies on organisations. Various governments have striven to maintain a controlling hand by pressing for such strategies as the use of 'key escrow' and of compulsory security hardware implants such as the 'clipper chip.' The implementation of 'big brother' controls is, depending upon your point of view, a worrying concern and certainly one that an organisation cannot afford to ignore.

14.1.2.4 Organization's Operatives

A secure network is only as secure as its weakest link, and history repeatedly points the finger at the human factor in an organisation as being just that. It is of no use whatsoever purchasing and installing the 'Rolls-Royce' of network security if one does not have the skilled and dedicated staff to operate it. An organisation must assume that competitors will make substantial and searching efforts to gain access to secrets that are valuable to them and will do so by exploiting any opportunity that presents itself. Whilst the majority of applications covered by this book have emphasised the cryptographic, hardware and software implications, it would be extremely short-sighted to adopt any of their features without appropriate consideration to the network operatives. Cryptanalysis has been described in terms of super-fast computers and expansive work forces that may be assembled at enormous expense with the sole objective of penetrating a target network. It is invariably the case whereby, for the price of a new car, a few thousand dollars, yen or pounds, or the mere crossing of the palm with silver, a network insider can deliver the goods to the 'third party'. The leakage can be achieved with greater efficiency than the spending of perhaps billions of dollars that organisations like the NSA sink into state-of-the-art computer attacks.

It is therefore imperative that a network administrator assemble a team of highly motivated and skilled personnel. The requirement is for a team that is both able and dedicated to the task of securing the network in the manner defined by the strategy agreed upon at the outset. Manpower training then becomes a major issue, as the failure to establish and maintain such a team will lead to management itself being the weakest link.

14.2 Infrastructure and Planning

It does not matter if the subject is football, marketing or scientific research: unless there is a plan of action, success will be reliant on fortune alone. This is not acceptable in the realm of high security, where far too much is at stake to rely upon chance. A formal plan can be defined as having three specific goals:

- Strategic goals
- Tactical goals
- Operational goals

14.2.1 Strategic Goals

These are broadly defined targets that are usually set by top management. The latter may have no technical appreciation of what steps must be taken to achieve these targets, but the results are seen to be greatly beneficial to the organisation. For example, a dignitary, head of state or member of a royal family might feel that their personal and official communications are vulnerable to attack and that if confidentiality were not ensured, serious repercussions for the monarchy might follow. That party will then establish a general goal that their communications must be secured, i.e. a strategic goal is set. In a military sense, we would expect the people responsible for making strategic goals to be generals or colonels. These are officers who very often have neither skills nor the time to attend to day-to-day security tasks. The action to realise that target then becomes a tactical goal and the responsibility of a different animal.

14.2.2 Tactical Goals

Tactical goals are the first step in achieving the strategic goals. They are usually set by middle management, e.g. captains, and are more specific about what must be achieved and are often more quantifiable. In the case of our aforementioned, threatened dignitary, a security subcommittee might decide that secure communications should be achieved by the use of encryption techniques and access controls. One would expect the members of this subcommittee to be more aware of the technological implications of their decision to the extent that they might make decisions about a provider or product. It is probably at this juncture in the proceedings that the security manager becomes involved for the first time. Their experience and advice would certainly be welcomed in setting the tactical targets.

14.2.3 Operational Goals

These goals are all about making it happen. They are highly specific in defining how the tactical goals are met and usually will be made by lower management, typically lieutenants in the military model or the security manager in a VIP or commercial institution. This body will be responsible for implementing the security hardware, network software and security parameter programming, day-to-day management and troubleshooting, key management and all that that entails, maintenance procedures and the control over the operators at the lowest levels of the hierarchy. The issues of establishing an operational hierarchy and personnel training become pertinent at this stage.

14.3 Operational Hierarchies

The establishment of an operational hierarchy is an important formal declaration about who is responsible for what and to whom each member of the team is responsible. There is nothing worse in an organisation, especially one responsible for security, than one in which the players are neither sure about their precise roles, nor sure about from whom they should be taking their orders. A well-defined chain of command, with a full task description at each level, instils confidence, a sense of team spirit and a sense of responsibility. The actual infrastructure will depend upon the specific nature of the network, the mission, the technology, or technologies involved, the geographical disposition and the environments discussed previously in this chapter. However, one might expect the situation to be rather like that portrayed in Figure 14.2. In smaller networks, it might well be the case that the security manager, network manager and key manager are all to be found in the same body. However, with communications so diverse and each involving their distinct technical know-how, it becomes increasingly more difficult for a security manager to take control of all aspects. What has become apparent from the author's experience is that during the inevitable rotation of personnel as staff depart to be replaced by newcomers, continuity of operations can suffer enormously. It is therefore essential that trained support staff be available to take over the reins of key players, should any of them cease to be available. The responsibilities of the operational security manager demand a very special executive and, as such, represent a post to grow into, not one to be selected haphazardly. This is very much the network-defining character, and its selection must be appropriate to the integrity and responsibility that it beholds.

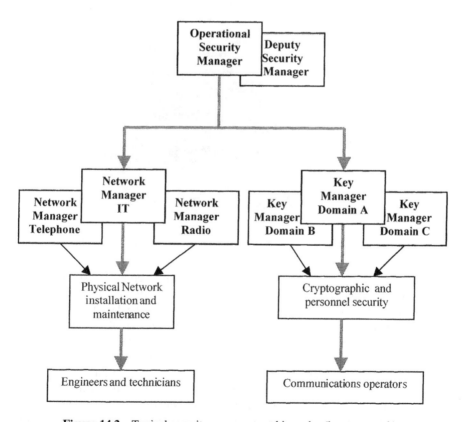

Figure 14.2   Typical security management hierarchy (large network)

14.4 Training

The Enigma machine (see Figure 14.3) that carried the secrets of Nazi Germany's secrets was undoubtedly a technical marvel in its day, despite the inherent weaknesses that became apparent with time. However, the greatest inroads into its operation were made as a result of operator errors. There are hosts of well-recorded incidents of operator failings that range from the basic set-up mistakes to downright carelessness and arrogance.

Operator training on the cipher machine did not require a complex procedure to be followed. The daily routine of the Enigma required the operator to select the correct rotors for the day and then set up the rings on the rotors to a predetermined position. Once the rotors were returned to the machine body, the user was instructed to spin them well to impart some randomness of the start position when it came to cipher or decipher a message. Before dealing with any message, however, the user had to choose, randomly it was supposed, a three-letter indicator for the first message of the day for synchronisation purposes.

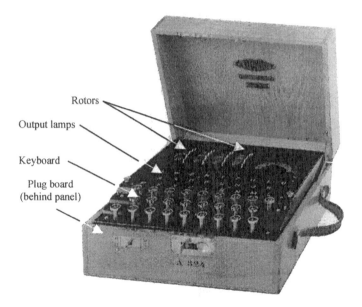

Rotors

Output lamps

Keyboard

Plug board
(behind panel)

Figure 14.3 The German WWII Enigma ciphering machine (four wheels)

Michael Smith, in his 'Station X', described how the British cryptanalysts at the Government Code & Cipher School (GCCS) at Bletchley Park during 1940–1945 postulated around this procedure to find a way in to determine the rotor setting for the day. The total number of possibilities available for the rotor set-up was 17,576, and testing all of these was a time-consuming task. Time was not what the GCCS had on their side, and so any method of reducing this brute-force attack effort would represent a welcome relief (remember the door pad model in Chapter 2?). Analysing the procedure, the GCCS team put themselves in the position of a lazy German operator, probably bored with routine and following the laid-down instructions with less attention than was required by his superiors. Then, instead of thinking of three random letters to enter, the operator simply chose the letters that existed as the rotors were entered back into the machine. In this action, the operator would be giving away the rotor set-up. A concentrated effort by the British eavesdroppers did in fact establish that often, the first messages of the day resulted in a spread around the Enigma daily rotor settings. This left the analysts with a mere 20–30 options to try instead of 17,576, in order to obtain the German set-up.

It also became apparent that in later variations of Enigma, when six start letters were to be entered at the start of the day, operators often failed to spin the rotors to achieve a random start point once the machine had been set-up. Instead of following management guidelines, operators unwittingly opened further niches for the British cryptanalysts to exploit. The all-important foothold was given by operators setting the start positions with less than random-ness, by inserting the names and initials of their loved ones Gün..Ter, Ren..Ate, or of Hit..Ler, of Lon..Don, etc. to such an extent that the allied listeners could easily identify individual operators.

One of the inherent failings of the Enigma was that it never coded a plain letter with the same ciphered letter. In other words, a plain 'A' would never be replaced by a cipher text 'A'. On its own, this was not perhaps a disastrous condition, but on one occasion, when a German operator was required to send a test message, he responded by simply pressing the letter 'L' and keeping their finger on the key to complete the entire message. This message was immediately recognised by the British eavesdropping team, for the ciphered message did not carry a single 'L', and so in that single instance, they had captured the wiring of a new rotor.

No doubt, there were occasions when the allies were as lax as the axis forces, but such occasions are certainly less well documented. In any case, no matter what the allegiance, the lack of discipline, which represents a failure in the training to iron out such operational flaws, often had a catastrophic consequence. To this day, training is very often treated with diffi-dence by both the purveyor of encryption devices and their client. When sales representatives are negotiating the price of a secure system, training courses are often the easiest components cut in length or to be eliminated altogether, to trim the budget. There is an understandable ploy by sales people to offer free training instead of giving a discount on every hardware unit sold and those sold in the future. Whilst this might be a good tactic as far as finance is concerned, it does give the impression that if training is offered 'free of charge', it is not really of much worth.

Where training is given, economy is still on the mind of system manager, and whereas courses are normally intended for a finite group of trainees, uninformed customer opinion is often that more can be achieved by more people attending. Of course, this is a very short-sighted view, and there is nothing more certain in training that as the class numbers increase, the information exchange decreases proportionally. As far as the numbers attending security courses are concerned, small is beautiful.

In many parts of the world, it is difficult to find personnel who have the ability, interest, and diligence as well as the strength of character to take on the role of security manager. In the author's experience, these senior staff are often very reluctant to take on the job in a profes-sional manner, and come the day when their training is complete and the manufacturer's support team has left, their inadequacies are exposed. The responsibility of the organisation's security then lies on unprepared shoulders. On too many occasions, knowledge and advice have been flowing up the management hierarchy, instead of down, and often the result is that junior staff are making tactical and operational decisions that are way beyond their official capacity. Now, the danger is that to capture the company secrets, it becomes easier and less expensive as the responsibility for security slips down the ladder, and the enemy only need supply a second hand Ford rather than a new Mercedes to get what they want.

There is a further role in training to be considered. It is an occasion when a manufacturer can profit from the extended contact and influence that training courses should provide. Great benefit can come from good-quality courses run by enthusiastic and skilled instructors and after-sales support teams.

14.5 Customer Support

It is false economy to adopt the 'sell and forget' stance that some manufacturers do engage, and clients should be well aware that when buying a security system after-sales support is paramount. The practicalities of installing a secure communications net are far removed from the neat and ordered presentations on the drawing board. Local problems at remote stations, interfacing with different systems and communications providers and environmental problems, all present obstacles to be cleared before a network can be deemed to be 'up and running'. Even then, new situations arise, current installations change, and new ones develop, and it is to everyone's long-term benefit that customer support is present long enough for the major problems to be solved and efficient, secure, working practice established. It is only then that the client can make the move to become totally secure by complete security separation from their supplier. To achieve this seamless transition, the security manager must be clear in his/her mind about the direction they should take and also have the confidence to take the difficult decisions that separation entails. Far too often, the author has returned to projects to find that security separation has not been completed because of various circumstances. Lack of confidence and knowledge on the part of the client's representative are common factors. Under these circumstances, the strategic and tactical goals will not have been met, just as in the case where the German High Command held implicit trust in its Enigma and its successful operation to achieve its goals. High ideals can be very far removed from reality, and security networks can be totally undermined by the weakest link. For the most part, that element is the human element.

14.6 Troubleshooting

It should be apparent by now that the role of the security manager involves many tasks including, as we have seen latterly, the training and discipline of his staff. Throughout the life of a network, problems will arise, and only constant monitoring can identify them and generate their solutions. So, finally, there follows a short presentation on the formalities of troubleshooting.

The first stage of problem solving is recognizing that a problem exists, that there is a discrepancy between the required state of affairs and the current situation. There are three stages to address this.

14.6.1 The Scanning Stage

This stage involves monitoring the network condition for any changing circumstances. At this time, the security manager may only vaguely be aware that there is a change to an environmental condition that might lead to a problem.

14.6.2 The Categorisation Stage

Here, an attempt is made to understand and verify that a discrepancy does exist. The manager should be attempting to formulate the type of problem, e.g. an equipment failure, a transmission medium failure or one of human error?

14.6.3 The Diagnostics Stage

This stage involves the gathering of information about the cause for concern and thereafter, specifying both the nature and the root of the problem. Once these factors have been established, one should be able to visualise what will be the likely outcome of the problem persisting for any length of time.

14.6.4 Generating Solutions

The second step in the decision-making process is to generate a number of alternative solutions, perhaps by a team brainstorming session, so that the best solution may be selected. Consider the situation where the problem is a poorly skilled operator; the first solution that might come to mind might be:

- Shoot them
- Sack them
- Demote them
- Fine them

However, after considerable thought, there might be less obvious solutions that would be more productive than the knee-jerk responses above: moving the operators to another task, training them to be a better operative or even, for the more devious, the possibility of using their failings to the organisation's benefit. Perhaps, it might be beneficial to allow his indiscretions to supply planted information to an eavesdropper.

Whatever final solution is arrived at, the following six criteria should be met:

- *Feasibility*: Can the solution be implemented within related organisational constraints such as, time, budget, technology and policy?
- *Quality*: How effectively can the solution solve the problem? Any alternatives that only offer partial solutions should be discarded.
- Acceptability: How will the organisation and, for that matter, the offending operator, be affected by the adopted solution, and is it therefore acceptable to all parties, or any?
- *Cost*: In security systems, this is perhaps easier to quantify than other applications. It is probably a straightforward task to compare the costs of letting the problem persist or in dealing with it in a conclusive manner.
- *Reversibility*: Clearly, executing our hapless operator is not a reversible option, but the substitution of a replacement cipher machine or a change in access controls would be possible to reverse if it failed to rectify the problem.

- *Ethics*: A sensitive manager might feel that the ethics of allowing the offending individual to be used as a 'Trojan horse', without their knowledge, would trouble the manager's conscience, and therefore this option would be negated. Conversely, if the consequences of an 'information leak' are serious enough, execution might put everyone at ease, in one way or another.

The final step in troubleshooting is having the ability and courage to take the required action in restoring the confidentiality, integrity and authenticity of the network. Choose your security manager well, for it is he/she who keeps your secrets secret.

04 33 63 89

References

Bartol, K.M. and Martin, D.C., '*Management*', Third edition, Irwin/McGraw-Hill, Boston, MA, 1998.

Hendry, M., '*Practical Computer Network Security*', Artech House, Norwood, MA, 1995.

Khan, D., '*The Codebreakers*', Sphere Books, London, 1973.

Oppliger, R., '*Internet and Intranet Security*', Artech House, Norwood, MA, 1997.

Purser, M., '*Secure Data Networking*', Artech House, Norwood, MA, 1993.

Redl, S.M., Weber, M.K. and Oliphant, M.W., '*An Introduction to GSM*', Artech House, Norwood, MA, 1995.

Schneier, B., '*Applied Cryptography*', Wiley, New York, 1996.

Smith, M., '*Station X*', Channel 4 Books, Macmillan, London, 1998.

Smith, M., '*The Emperor's Codes*', Bantam Press, Transworld Publishers, London, 2000.

Smith, R., 'Deciphering the advanced encryption standard', *Network Magazine*, 2001.

Index

Access control, 59, 69, 100, 149, 241, 243, 253–256, 264, 294, 308
Access Tokens, 14, 101, 256
Adaptive frequency, 181
Adaptive power transmission, 173, 185
Address resolution protocol, 282
Administration layer, 214, 221, 223
Advanced encryption standard, 43, 58, 97, 139, 304
AES (see Advanced encryption standard)
Algorithm A3, 118–125, 141
Algorithm A5, 118–126, 141
Algorithm A8, 118–126
Algorithms, 38, 58, 184, 201, 202, 305
  asymmetrical, 6, 7, 9, 38, 305
  classes, 38
  multi, 99
  proprietary, 69, 83, 98, 305
  symmetrical, 38, 305
Analogue encryption, 26, 62
Analysis of residual plain information, 243, 251
Army data transmission, 292
ARQ, 290
Asymmetric encryption, 6, 7, 272, 273
Asymmetrical cryptography, 38–40
Authentication, 2, 4–6, 38, 76, 88, 100, 120, 135, 147, 255, 271, 314
AuC (see authentication centre), 114–119
Authentication centre, 116, 119

Automatic key selection, 230, 232
Availability, 10

Battlefield link communications, 222
Biometric access tools, 13, 253, 254
Biometrics, 13–16, 254, 304
Black
  circuit, 20
  designation, 16
Bletchley Park, 11, 304, 310
Block ciphering, 33
Break-in, 187, 193, 196, 203
Brute force attack, 11, 126, 243, 250, 310
Bugging, 88
Bulk communications, 211
Bulk encryption, 211, 217, 224
Burn, 246, 294
Burst transmission, 177, 178

Capacitive coupling, 19
Cascade, 166–167
CBC (see Cipher block chaining)
CFB (see Cipher feedback mode)
Challenge/response control, 13, 14, 119, 270
Chip cards, 14, 49, 70, 83, 98, 100, 103, 112, 116, 138, 162, 214, 220, 221, 253–255
Cipher
  block chaining, 33, 35, 247
  feedback mode, 36
  module architecture, 85

module, 84
   stream, 31
Clients, 306
Clipper chip, 307
Codebreakers The, 3, 26, 301
Co-location, 184
Colossus, 304
Combat net radio, 296
Compromise of information, 243, 251
Compromising emanations, 16, 101, 306
COMSEC, 182–202
Confidentiality, 1, 2, 7, 9, 88, 146–147, 164,
   243, 305, 308, 314
Countermeasures, 4, 18, 147, 171, 173,
   271
Cross coupling, 17
Cryptanalysis, 1, 3, 89, 242, 302, 307
Cryptographic design philosophy, 289
Cryptography, 3, 38–39, 47, 55, 271, 304
Cryptology, 3, 43, 62

Data communications, 289
Data courier device, 203
Data encryption, 135, 217, 221, 245, 295
Data encryption standard, 38, 43, 47, 58, 97,
   135, 305
Data over radio links, 290
Deciphering of encrypted file, 247
Default keys, 76, 135, 164
DEFNET, 235, 236
Denial of service, 88, 269
DES (see Data encryption standard)
Dictionary attack, 12
Diffie–Hellman algorithm, 42, 305
Digital
   ciphering, 30, 70
   encryption device, 72
   scrambling, 67, 68
   signature algorithms, 41
   signatures, 8–10, 40–42, 271–272
   stream ciphering, 31
Direct sequence encoding, 177–180
Disk block, 253
Disk encryption, 245, 251
DLL (see Downline loading)
DoS (see Denial of service)

Downline loading, 77, 98, 103
Dynamic co-ordination, 199

EASYNET, 233
Eavesdropping, 1, 4, 26, 61, 63, 87, 122,
   147, 242, 249, 265, 313
ECB (see Electronic code book)
ECCM (see Electronic counter
   countermeasures)
Echelon, 265
Economic element, The, 304
Electronic attack, 171
Electronic code book, 33, 34
Electronic counter countermeasures, 171,
   306
Electronic protection measures, 171
Electronic support measures, 171
Electronic warfare, 171
Element
   economic, 304
   international, 305
   legal/political, 305
   sociological, 306
   technical, 304
E-mail, 10, 13, 265–274
   monitoring, 265
   scenario, 265
   security guidelines, 274
EMC protection, 18, 22, 23, 65, 88–89, 101,
   243–244, 295
Emperors' keys – The, 3
Encrypted radio communications, 146
Encryption, 2, 4–9, 13–16, 26–27, 30–51,
   54–57, 62, 77, 80, 83, 89, 96
   telephone, 87
Enhanced digital radios, 296
Enigma, 1, 2, 11, 304, 310–312
Environments, 303
   global, 303
   local/task, 306
EPM (see Electronic protection measures)
EPM (see Electronic protection methods)
Ethics, 314
Eurolink, 216, 218, 223
Evaluation of encryption equipment, 57
EW (see Electronic warfare)

Exhaustive key search, 11, 44, 124, 243
Export controls, 306

Fallback keys, 83, 244, 252, 296
Fax machines, 228
    operating parameters, 239
Fax management centre, 237
FEC (see Forward error correction)
File encryption, 242, 246
Fill gun, 83, 98, 103, 218
Fingerprints, 11, 13–14
Firewall, 257
Forward error correction, 73, 74, 291
Frame alignment signal, 212
Frequency
    hopping-slow, 127
    management, 145, 149, 162–163,
        185–186
    scrambling, 28
Frequency hopping, 71, 117, 118, 121, 127,
    171, 177–179, 191, 194

Galvanic coupling, 20
GAN (see Global area networks)
General packet radio system, 139–141
Global area networks, 91, 95, 239, 291, 293,
    304
Global environment, The, 291, 303, 306
Goals
    operational, 308
    strategic, 308
    tactical, 308
Government Code & Cipher School, 310
GPRS (see General packet radio system)
GSM
    cellular system, 113, 114
    custom security, 128–133, 137
    data channel, 129
    encryption, 124
    hand phone security module, 136
    monitoring, 123, 127
    security, 115–118, 124,
GSM algorithms, 114, 123

Hailing, 187, 191, 197–199, 203
Hailing hops, 197, 198

Hardware security features, 101
Hash algorithms, 38, 41
Hash function, 7, 15, 41–42, 272
Have quick, 206
HF radio installation, 63
HLR (see Home location register)
Home location register, 115–141
Hopping rate, 181, 184–186, 195, 208
Hopping sequence, 182–187
Hybrid encryption, 274

IDEA, 58
Illegal access, 252
Illegal boot up, 256
IMSI, 117–121
Inductive coupling, 20
Information disclosure, 267, 268
Initialisation vector, 32, 74, 133
INMARSAT communications, 91, 94, 240,
    304
Integrity, 1, 2, 8, 9, 146, 169, 242, 248, 269,
    303, 309, 314
International element, The, 305
Internet, 253, 257, 265, 275, 292, 298, 299
Internet worm, 269
IP security protocol, 277
ISAKMP/Oakley, 277
IV (see Initialisation vector)

Jamming, 89, 171–183, 203
    broadband, 172, 176, 178
    combined, 176, 183, 184
    pulse, 172, 176
    self, 176, 184–185, 198, 199
    swept, 172, 176

Kerckhoff's Assumption, 124
Key
    agreement algorithms, 42
    changes, 1–2, 51, 54, 166, 262, 263
    destruction, 54
    distribution, 48, 70, 71, 77, 150, 219,
        243, 260, 263
    escrow, 307
    generation, 45, 243
    generators, 73–75, 133

lifetime, 45
management centre, 59, 82, 214, 237
management, 44, 59, 70, 71, 77, 83, 99,
    165, 251, 295, 308
signature, 46, 81, 100, 132, 138, 262, 295
storage, 47, 77, 135
transport devices, 49
transport key, 103, 162, 214, 255
variety, 11, 83
Key to channel assignment, 85, 163
Keystream, 31, 73, 75, 192, 216
    period, 56, 69, 83 135
Ki, 127

LAN security, 256, 267, 299
LAN workstation scenario, 257
Laptop field terminals, 292, 293
Late entry, 68, 83, 194–196
Layer 2 tunnelling protocol, 277
LCM (see Long cycle mode)
Legal, political element, The, 305
Link encryption, 211
Long cycle mode, 33, 37
Long term key, 263
Loss of data, 244, 294, 295

MAC (see Message authentication code)
Mail server, 265
Manual key selection, 232
MARS, 43
Masquerading, 4, 242, 249, 267, 269, 270
Message authentication code, 6, 7, 38, 41,
    247–249
Milink, 223, 225
Military battlefield communications, 289
Military communications, 290
Military voice communications, 289
Modification, 267, 269
Modulated harmonics, 16
Modulo 2 addition, 73, 214
Multi-key networks, 233–234, 237

Naval communications, 63
Net entry, 190, 193, 196, 197, 207

Network
    capturing, 270
    data service, 299
    manager, 146, 150
Non-repudiation, 9
Notebook scenario, 258
NSA, 305, 307
Null volts plane, 21

One way function, 47
Operational hierarchies, 308
Orthogonal, 178, 185
OTAR (see Over the air re-keying)
Over the air re-keying, 77, 98, 156, 166, 289

Packet
    data communications, 289
    radio techniques, 297
    radios, 296
Packet filtering, 283–284
Passwords, 10–14, 59, 77, 100, 242–245,
    249–253, 264, 265
    guidelines for use, 12
Password cracker, 12
PC security organisation, 259
PCMCIA cards, 48, 240, 244, 253, 256,
    264, 294
Pearl Harbour, 301
PGP (see Pretty good privacy)
Phonemes, 25, 67
PINs, 10, 14, 249, 253–255
Plain override, 69, 83, 168
Policy, 178, 271
Pretty good privacy, 10, 11, 40, 241
Private key, 6, 39, 97, 221, 272
Private routing table, 282
Process gain, 181, 184, 198
PSTN, 87, 114, 143–164
PSTN ciphered call set up, 90,
Public key, 6, 9, 39, 97, 221, 272, 273
Public routing table, 282

Radio cipher module, 160
Radio networks, 145
Random key selection, 235
RARASE, 75, 76

RC6, 43
Red
    circuit, 20
    designation, 16
Red/black separation, 18, 63, 89, 96, 101,
    136, 256
Remote blocking, 168
Remote key cancelling, 168
Replay, 4, 83, 239, 267, 269
R-FEC (see Forward error correction)
Rijndael, 43
RSA, 6, 8, 38, 40, 305

Scrambling, 25
    frequency, 28
    time element, 28
Secure e-mail, 241
Security
    administration, 262, 306–7
    manager, 3, 25, 100, 108, 150, 250, 303,
        306–309, 311, 314
    modules, 15, 18, 243, 245
    threats to PCs, 241–244
Separation, 55, 57, 61
Serpent, 43
Session key, 76, 133–134, 214, 242,
    246–247, 273, 295
Short message service, 297
Silent mode tracking, 168
SIM card, 116–126, 141
Smart card (see chip card)
Sociological element, 306
Software encryption, 241
Solutions, 313
Spoofing, 4, 14, 69, 89, 249, 267, 269, 270
Spread spectrum techniques, 175, 177
Static co-ordination, 198
Station X, 3, 310
STU, 92, 94–98
Suppliers, 306
Symmetric encryption, 7, 68, 273
Symmetric session key, 134
Symmetrical encryption, 14, 38, 39, 221
Synch. bursting, 69
Synchronisation, 59, 67, 68, 73, 76, 99, 132,
    149, 156, 160, 178, 191, 194, 207, 310

Tamper proofing algorithm, 135
Tamper resistance, 99, 244
Tamperproof, 15, 59, 250
    key, 106, 135, 244
    PC card, 250,
Tamperproof security modules, 99, 129,
    250
TCP/IP, 299
Technical element, 304
Telephone
    ciphering device, 98
    patch, 152, 153, 160–165
    secure, architecture, 102
    security, 87, 89, 101
TEMPEST, 16, 18, 21, 65, 88, 227,
    243–244, 295
Thales, 156
Threats, 4, 88, 112, 146, 241, 244, 267,
    284–285
Three key system, 51–53,
Time authentication, 5, 6, 69, 76, 83, 147,
    163, 168
Time element ciphering, 28
Time master flyby, 207
TMSI, 116–121
Training, 59, 271, 301, 307, 308, 310–312
TRANSEC, 182–194
Transparent channel, 129
Transport mode, 277
Trojan Horse, 18, 249, 269, 314
Troubleshooting, 312, 314
    categorization stage, 313
    diagnostic stage, 313
    generating solutions, 313
    scanning stage, 312
Trunking systems, 177
Tunnel table, 282
Tunnelling, 276
Twofish, 43

Ultra, 2
Unauthorised
    intrusion, 244, 252
    manipulation of data, 242–244, 247, 249,
        294
    read out, 242, 244–246

user, 12, 245, 270, 271
   access, 12, 147, 242, 244, 249, 294
US export regulations, 305
User groups, 273
VECTOR, 156,
VIPNET, 143, 146–165
Virtual memory, 251
Virtual private network
   definition, 275
   example application, 285–287
   packet header formats, 278–281
   Security association list, 281–282
   tunnel table, 282

Virus, 241, 258, 269, 271
Virus protection, 274
Visitor location register, 116–120, 127
VLR (see Visitor location register)
Voice communications, 61
VPN (see Virtual private network)

WAN, 256
Wassenaar agreement, 305
Wipe, 241, 246, 294
Wiped, 251

Yamamoto, Admiral Isoroku, 2, 301, 302

Printed and bound in the UK by
CPI Antony Rowe, Eastbourne

Printed and bound by CPI Group (UK) Ltd, Croydon, CR0 4YY

27/10/2024

14580217-0002